The Water Footprint Assessment Manual

The Water Footprint Assessment Manual

Setting the Global Standard

Arjen Y. Hoekstra, Ashok K. Chapagain,
Maite M. Aldaya and Mesfin M. Mekonnen

Routledge
Taylor & Francis Group
LONDON AND NEW YORK

earthscan
from Routledge

First published in paperback 2024

First published by Earthscan in the UK and USA in 2011

For a full list of publications please contact:
Earthscan
4 Park Square, Milton Park, Abingdon, Oxon OX14 4RN

and by Routledge
605 Third Avenue, New York, NY 10158

Earthscan is an imprint of the Taylor & Francis Group, an informa business

Publisher's Note
The publisher has gone to great lengths to ensure the quality of this reprint but points out that some imperfections in the original copies may be apparent.

A catalogue record for this book is available from the British Library

Library of Congress Cataloging-in-Publication Data

The water footprint assessment manual : setting the global standard / Arjen Y. Hoekstra ... [et al.].
p. cm.
Includes bibliographical references and index.
ISBN 978-1-84971-279-8 (hardback)
1. Water consumption–Measurement. 2. Water consumption–Environmental aspects. 3. Water-supply–Accounting. I. Hoekstra, Arjen Y., 1967–
TD499.W384 2011
333.91'13–dc22

2010047901

ISBN 13: 978-1-84971-279-8 (hbk)
ISBN 13: 978-1-03-292776-3 (pbk)
ISBN 13: 978-1-84977-552-6 (ebk)

DOI: 10.4324/9781849775526

Typeset by JS Typesetting Ltd, Porthcawl, Mid Glamorgan

Cover design by Rob Watts; water footprint design by Angela Morelli

Contents

List of Figures, Tables and Boxes ix
Acknowledgements xiii
Preface xvii
Acronyms xix

1 Introduction **1**
1.1 Background 1
1.2 The water footprint concept 2
1.3 Water footprint assessment 3
1.4 Guide for the reader 5

2 Goals and Scope of Water Footprint Assessment **7**
2.1 Goals of water footprint assessment 7
2.2 Scope of water footprint accounting 9
2.3 Scope of water footprint sustainability assessment 15
2.4 Scope of water footprint response formulation 16

3 Water Footprint Accounting **19**
3.1 Human appropriation of fresh water: What do we measure and why? 19
3.2 Coherence between different sorts of water footprint accounts 21
3.3 Water footprint of a process step 23
3.3.1 Blue water footprint 23
3.3.2 Green water footprint 29
3.3.3 Grey water footprint 30
3.3.4 Calculation of the green, blue and grey water footprint of growing a crop or tree 40
3.4 Water footprint of a product 46
3.4.1 Definition 46
3.4.2 Schematization of the production system into process steps 47
3.4.3 Calculation of a product water footprint 48
3.5 Water footprint of a consumer or group of consumers 52

3.5.1 Definition 52
3.5.2 Calculation 52
3.6 Water footprint within a geographically delineated area 53
3.6.1 Definition 53
3.6.2 Calculation 53
3.7 National water footprint accounting 54
3.7.1 The national water footprint accounting scheme 54
3.7.2 Calculation of the water footprint within a nation 55
3.7.3 Calculation of the water footprint of national consumption 56
3.7.4 Water savings related to trade 60
3.7.5 National water dependency versus water self-sufficiency 61
3.8 Water footprint accounting for catchments and river basins 61
3.9 Water footprint accounting for municipalities, provinces or other administrative units 63
3.10 Water footprint of a business 63
3.10.1 Definition 63
3.10.2 Choosing the organizational boundaries of the business 65
3.10.3 Calculation of the business water footprint 68

4 Water Footprint Sustainability Assessment **73**
4.1 Introduction 73
4.2 Geographic sustainability: Sustainability of the water footprint within a catchment or river basin 76
4.2.1 Introduction 76
4.2.2 Environmental sustainability criteria for identifying environmental hotspots 78
4.2.3 Social sustainability criteria for identifying social hotspots 87
4.2.4 Economic sustainability criteria for identifying economic hotspots 88
4.2.5 Assessing primary and secondary impacts in the hotspots identified 88
4.3 Sustainability of the water footprint of a process 89
4.4 Sustainability of the water footprint of a product 91
4.4.1 Identifying the unsustainable components in the water footprint of a product 91
4.4.2 Water footprint impact indices reflecting local environmental impacts 94
4.5 Sustainability of the water footprint of a business 97
4.6 Sustainability of the water footprint of a consumer 97

5 Library of Water Footprint Response Options **99**
5.1 Shared responsibility 99
5.2 Reducing the water footprint of humanity: What is possible? 99
5.3 Consumers 103
5.4 Companies 106
5.5 Farmers 107
5.6 Investors 109
5.7 Governments 110

6 Limitations **115**

7 Future Challenges **119**
7.1 Water footprint assessment methodology and data 119
7.2 Application of the water footprint in different contexts 122
7.3 Embedding the water footprint in existing water and environmental accounts and reports 123
7.4 Linking to ecological, energy and carbon footprint methods 124
7.5 Linking to material flow analysis, input-output modelling and life cycle assessment 125

8 Conclusion **129**

Appendix I Calculation of Green and Blue Evapotranspiration Using the CROPWAT Model *131*
Appendix II Calculating the Process Water Footprint of Growing a Crop: An Example for Sugar Beet in Valladolid (Spain) *135*
Appendix III Calculating the Water Footprint of a Product: Example for Refined Sugar from Valladolid (Spain) *143*
Appendix IV Examples of Grey Water Footprint Calculations *147*
Appendix V Environmental Flow Requirements *151*
Appendix VI Frequently Asked Questions *155*

References *169*
List of Symbols *183*
Glossary *187*
Index *199*

List of Figures, Tables and Boxes

Figures

1.1 Schematic representation of the components of a water footprint. It shows that the non-consumptive part of water withdrawals (the return flow) is not part of the water footprint. It also shows that, contrary to the measure of 'water withdrawal', the 'water footprint' includes green and grey water and the indirect water-use component 3
1.2 Four distinct phases in water footprint assessment 4
3.1 The green and blue water footprint in relation to the water balance of a catchment area 20
3.2 Process water footprints as the basic building block for all other water footprints 22
3.3 The direct and indirect water footprint in each stage of the supply chain of an animal product 24
3.4 The relation between the water footprint of national consumption and the water footprint within a nation in a simplified example for two trading nations 24
3.5 Blue water footprint accounting in the case of water recycling and reuse 28
3.6 The subsequent processes in irrigation: storing water, transport of water, irrigation on the field. Each process step has its own water footprint 43
3.7 Schematization of the production system to produce product *p* into *k* process steps. Some steps are in series, others are parallel. The water footprint of output product *p* is calculated as the sum of the process water footprints of the processes that constitute the production system. Note: this simplified scheme presupposes that *p* is the only output product following from the production system 48
3.8 Schematization of the last process step in the production system to produce product *p*. The water footprint of output product *p* is calculated based on the water footprints of the input products and the process water footprint when processing the inputs into the outputs 49

3.9 The national water footprint accounting scheme. The accounting scheme shows the various balances that hold for the water footprint related to national consumption ($WF_{cons,nat}$), the water footprint within the area of the nation ($WF_{area,nat}$), the total virtual-water export (V_e) and the total virtual-water import (V_i) 56
3.10 The catchment water footprint accounting scheme. The accounting scheme shows the various balances that hold for the water footprint of consumers living within the catchment, the water footprint within the catchment area, the total virtual-water export from the catchment and the total virtual-water import into the catchment 62
3.11 Composition of the water footprint of a business 64
3.12 Business that consists of three business units producing products A–C. Product inflow $I_u[x,i]$ refers to the annual volume of input product i from source x into business unit u. Product outflow $P_u[p]$ refers to the annual volume of output product p from business unit u. Product flow $P^*_u[p]$ refers to the part of $P_u[p]$ that goes to another business unit within the same business 68
4.1 Assessment of the sustainability of the water footprint within a catchment or river basin in four steps 76
4.2 The blue water footprint over a year compared to the blue water availability, where the latter is equal to run-off (under undeveloped conditions) minus environmental flow requirements 83
II.1 Climate station in Valladolid (Spain) (dot in black) and sugar beet harvested area in Spain (unit: proportion of grid cell area) 136
III.1 Spanish refined sugar production (from sugar beet) diagram including product fractions 144

Tables

2.1 Spatiotemporal explication in water footprint accounting 12
3.1 Examples of the components of a business water footprint 64
4.1 Example of how to assess the extent to which the water footprint of a product is sustainable, based on two criteria: geographic sustainability of the water footprints in the catchments in which the process steps are located and sustainability of the underlying process steps themselves. Priority components in the water footprint of a product can be identified based on which components are unsustainable and the share of a component in the total water footprint of the product. The table needs to be filled separately for the green, blue and grey water footprint of the product 93

5.1 Possible water footprint reduction targets per sector and water footprint component 101
5.2 Priorities in water footprint reduction 103
5.3 Corporate water footprint response options 108
5.4 Options for crop farmers to reduce their water footprint 109
5.5 Options for governments to reduce water footprints and mitigate related impacts 111
7.1 An overview of water footprint studies 122
7.2 How water footprint assessments can feed LCA 126
II.1 Planting and harvesting dates and yield for sugar beet production in Valladolid (Spain) 136
II.2 Total green-blue water evapotranspiration based on the CWR output table of CROPWAT 8.0 137
II.3 Irrigation schedule under the rain-fed scenario: Output table of CROPWAT 8.0 139
II.4 Irrigation schedule under the irrigation scenario: Output table of CROPWAT 8.0 140
II.5 Calculation of the green and blue components of the process water footprint (m^3/ton) for sugar beet in Valladolid (Spain) using the CWR-option and irrigation schedule option for a medium soil 141
II.6 Calculation of the grey component of the process water footprint (m^3/ton) for sugar beet in Valladolid (Spain) 142
III.1 Green, blue and grey water footprint for sugar beet in Valladolid (Spain) (m^3/ton) 146

Boxes

2.1 Goals of water footprint assessment 8
2.2 Are there 'scopes' in water footprint accounting as there are in the case of corporate carbon footprint accounting? 14
3.1 The relation between the different sorts of water footprints 23
3.2 Unit of a water footprint 25
3.3 Data sources for the calculation of a blue water footprint 27
3.4 The history of the grey water footprint concept 31
3.5 The concept of critical load 34
3.6 The grey water footprint in different cases of point-source pollution 36
3.7 Three-tier approach in estimating diffuse pollution loads 38
3.8 Data sources for the calculation of the water footprint of 'growing a crop' 44

3.9 Terminology: Water footprint, virtual-water content, embedded water 46
3.10 What is new for companies when considering their business water footprint? 66
4.1 History of water footprint sustainability assessment 74
4.2 Sustainability criteria for water use and allocation within a catchment or river basin 77
4.3 Environmental green water requirement 81
4.4 The effect of the green water footprint on blue water availability 82
4.5 The sustainability of a blue water footprint depends on how it affects both blue-water flows and stocks 84
4.6 How the 'blue water scarcity' as defined in water footprint studies differs from conventional water scarcity indicators 86
5.1 Water neutrality 104
5.2 Water footprint offsetting 105

Acknowledgements

This manual has been written with the great help of many organizations and individuals. First of all we would like to thank all partners of the Water Footprint Network that have contributed in so many different ways to the maturing of the water footprint concept. We thank the following 130 organizations, all partners of the network (as per 16 October 2010): ADAS (UK), Adecagua (Spain), Allenare Consultores (Mexico), Alliance for Water Stewardship (US/Australia), AmBev – Companhia de Bebidas das Americas (Brazil), APESA (France), Arup (UK), Association du Flocon à la Vague (France), ATA – Ativos Técnicos e Ambientais (Brazil), Austrian Institute of Technology (Austria), Barilla (Italy), Beijing Forestry University (China), Bianconi Consulting (UK), Bionova (Finland), Blonk Milieu Advies (Netherlands), C&A (Germany), CEIGRAM – Research Centre for the Management of Agricultural and Environmental Risks, Technical University of Madrid (Spain), CESTRAS – Centro de Estudos e Estratégias para a Sustentabilidade (Portugal), Climate Change Commission (Philippines), Coca-Cola Hellenic (Greece), Confederation of European Paper Industries (Belgium), Consejo Consultivo del Agua (Mexico), Conservation International (US), CREM (Netherlands), CSE Centre for Sustainability and Excellence (Greece), CSQA Certificazioni (Italy), Cyprus University of Technology (Cyprus), Decide Soluciones Estratégicas (Mexico), Denkstatt (Austria), DHV (Netherlands), Directorate-General for Water Affairs (Netherlands), Dole Food Company (US), Eawag – Swiss Federal Institute of Aquatic Science and Technology (Switzerland), Ecolife (Belgium), Ecologic – Institute for International and European Environmental Policy (Germany), Ecological Society for Eastern Africa (Kenya), Ecometrica (UK), EcosSistemas Sustainable Solutions (Brazil), EMWIS – Euro-Mediterranean Information System on know-how in the Water sector (France), Enzen Water (UK), EPAL – Empresa Portuguesa de Aguas Livres (Portugal), Fibria Celulose (Brazil), First Climate (Germany), FloraHolland (Netherlands), Food and Drink Federation (UK), Fundación Centro de las Nuevas Tecnologías del Agua (CENTA) (Spain), Fundación Chile (Chile), Geoklock – Consultoria e engenharia ambiental (Brazil), Global Footprint Network (US), GRACE (US), Green Solutions (Chile), Grontmij (Netherlands), Heineken (Netherlands), iMdea Water Foundation (Spain),

Institut für Nachhaltige Landbewirtschaftung (Germany), International Finance Corporation (US), International Water Management Institute (Sri Lanka), Jain Irrigation Systems (India), Jutexpo (UK), Kingston University (UK), KWR – Watercycle Research Institute (Netherlands), Lafarge (France), Leibniz Institute for Agricultural Engineering Potsdam-Bornim (Germany), LimnoTech (US), Live Earth (US), Marcelino Botín Foundation – The Water Observatory (Spain), Massey University – Soil and Earth Sciences Group (New Zealand), McCain Alimentaire (France), Michigan Technological University – Center for Water and Society (US), National Ground Water Association (US), National University of Cordoba (Argentina), Natura Cosméticos (Brazil), Nestlé (Switzerland), Netherlands Water Partnership (Netherlands), Next Planet ASBL (Belgium), Oranjewoud (Netherlands), Pacific Institute for Studies in Development, Environment, and Security (US), Partners for Innovation (Netherlands), PE International (Germany), People 4 Earth (Netherlands), PepsiCo (USA), Plant and Food Research (New Zealand), PRé Consultants (Netherlands), PricewaterhouseCoopers, Province of Overijssel (Netherlands), PTS – Papiertechnische Stiftung (Germany), Pyramid Sustainable Resource Developers (Australia), Quantis (Switzerland), Química del Campo (Chile), Raisio (Finland), Redevco (Netherlands), Renault (France), RodaxAgro (Greece), Royal Haskoning (Netherlands), SABMiller (UK), Safe Drinking Water Foundation (Canada), SERI – Sustainable Europe Research Institute (Austria), Smart Approved WaterMark (Australia), Soil & More International (Netherlands), Source 44 (US), Stora Enso (Sweden), Summa Environmental Technologies (Ecuador), Swiss Development Agency (Switzerland), The Coca-Cola Company (US), The Nature Conservancy (US), Tobco (Belgium), UNEP (France), UNESCO-IHE Institute for Water Education (Netherlands), Unilever (UK), University of Chile (Chile), University of Natural Resources and Applied Life Sciences (Austria), University of São Paulo – Escola de Engenharia de São Carlos (Brazil), University of São Paulo – GovÁgua (Brazil), University of Siena (Italy), University of Tokyo (Japan), University of Twente (Netherlands), University of Zaragoza (Spain), UPM-Kymmene Corporation (Finland), URS Corporation (UK), USAID – United States Agency for International Development (US), Vewin – the Dutch Association of Drinking Water Companies (Netherlands), Viña Concha y Toro (Chile), Viña De Martino (Chile), Viña Errazuriz (Chile), Water Neutral Foundation (South Africa), Water Strategies (UK), Wildlife Trust (US), World Business Council for Sustainable Development (Switzerland), WWF – the global conservation organization (Switzerland) and Zero Emissions Technologies (Spain).

We thank the members of WFN's grey water footprint working group, which critically reviewed the grey water footprint concept and provided valuable suggestions for improving the definition and guidelines: Jose Albiac (CITA,

Spain), Maite Aldaya (University of Twente, the Netherlands), Brent Clothier (Plant and Food Research, New Zealand), James Dabrowski (CSIRO, South Africa), Liese Dallbauman (Pepsi, UK), Axel Dourojeanni (Fundación Chile, Chile), Piet Filet (WWF, Australia), Arjen Hoekstra (University of Twente, the Netherlands), Mark Huijbregts (Radboud University, the Netherlands), Marianela Jiménez (Nestlé, Switzerland), Greg Koch (The Coca Cola Company, US), Marco Mensink (CEPI, Belgium), Angel de Miguel García (IMDEA Agua, Spain), Jason Morrison (Pacific Institute, US), Juan Ramon Candia (Fundación Chile, Chile), Todd Redder (Limnotech, US), Jens Rupp (Coke Hellenic, Greece), Ranvir Singh (Massey University, New Zealand), Alistair Wyness (URS Corporation, UK), Erika Zarate (WFN, the Netherlands), Matthias Zessner (Vienna University of Technology, Austria) and Guoping Zhang (WFN, the Netherlands).

A second working group of the Water Footprint Network critically reviewed and proposed improvements to the method of water footprint sustainability assessment. We are grateful to all its members: Maite Aldaya (University of Twente, the Netherlands), Upali Amarasinghe (IWMI, Sri Lanka), Fatima Bertran (Denkstatt, Austria), Sabrina Birner (IFC, US), Anne-Leonore Boffi (WBCSD, Switzerland), Emma Clarke (Pepsi, UK), Joe DePinto (Limnotech, US), Roland Fehringer (Denkstatt, Austria), Carlo Galli (Nestlé, Switzerland), Alberto Garrido (Technical University of Madrid, Spain), Arjen Hoekstra (University of Twente, the Netherlands), Denise Knight (Coca-Cola, US), Junguo Liu (Beijing Forestry University, China), Michael McClain (UNESCO-IHE, Netherlands), Marco Mensink (CEPI, Belgium), Jay O'Keeffe (UNESCO-IHE, Netherlands), Stuart Orr (WWF, Switzerland), Brian Richter (TNC, US), Hong Yang (EAWAG, Switzerland) and Erika Zarate (WFN, Netherlands).

We also thank the members of the Scientific Peer Review Committee, who reviewed the draft of this manual: Huub Savenije (Delft University of Technology, the Netherlands), Alberto Garrido (Technical University of Madrid, Spain), Junguo Liu (Beijing Forestry University, China), Johan Rockström (Stockholm University & Stockholm Environment Institute, Sweden), Pasquale Steduto (FAO, Italy), and Mathis Wackernagel (Global Footprint Network, US). In addition, we thank Brian Richter (TNC, US) for reviewing a draft of the chapter on sustainability assessment.

There have been many other valuable inputs. We cannot mention the hundreds of individuals and organizations that have contributed by providing feedbacks on the water footprint concept and application by means of email and personal contact. We would like to mention, however, at least: the Food and Agriculture Organization of the United Nations, in particular Giovanni Muñoz, for valuable advice on the CROPWAT model; the World Bank Institute, particularly Mei Xie, for cooperating in the development of various water footprint training materials; the World Business Council for Sustainable Development for organizing a valuable workshop on the water footprint in Montreux, Switzerland, March 2010; the Beverage Industry Environmental Roundtable (BIER) for looking into the specific implications of the water footprint for the beverage sector; and Soil & More International for providing extensive feedback on the influence of soil management on the water footprint of crop production.

We thank the employers of the authors for allowing them to dedicate time to prepare and write the manual: University of Twente, employer of Arjen Hoekstra and Mesfin Mekonnen and former employer of Maite Aldaya; WWF-UK, employer of Ashok Chapagain; the Research Centre for the Management of Agricultural and Environmental Risks (CEIGRAM) of the Technical University of Madrid, former employer of Maite Aldaya; and the United Nations Environment Programme (UNEP), current employer of Maite Aldaya.

Finally, we thank the staff of the Water Footprint Network for their continued dedication, their contributions to the advancement of water footprint thinking, application and dissemination and for their friendship: Derk Kuiper, Erika Zarate and Guoping Zhang. Thanks to Joshua Waweru and Joke Meijer-Lentelink for their secretarial support and to René Buijsrogge for his help in maintaining the water footprint website.

Preface

This book contains the global standard for 'water footprint assessment' as developed and maintained by the Water Footprint Network (WFN). It covers a comprehensive set of definitions and methods for water footprint accounting. It shows how water footprints are calculated for individual processes and products, as well as for consumers, nations and businesses. It also includes methods for water footprint sustainability assessment and a library of water footprint response options.

A shared standard on definitions and calculation methods is crucial given the rapidly growing interest in companies and governments to use water footprint accounts as a basis for formulating sustainable water strategies and policies.

This manual has been prepared by the authors as requested by the WFN. The current manual is an updated, revised and expanded version of *Water Footprint Manual: State of the Art 2009*, published by the WFN in November 2009 (Hoekstra et al, 2009a). This new edition has been produced after intensive consultations with partners and researchers worldwide. Directly following the publication of the *Water Footprint Manual*, all partners of the WFN were invited to provide feedback on the manual. In addition, two working groups were formed, consisting of individuals from partners of the WFN and invited experts. One working group addressed questions around the grey water footprint (Zarate, 2010a); the other one studied issues pertaining to water footprint sustainability assessment (Zarate, 2010b). In addition, a number of partners initiated pilot projects in collaboration with the WFN that aimed at exploring the practical implications of using the water footprint in formulating a corporate water strategy or water policy in a specific geographical setting. On the basis of feedbacks received – new scientific publications, experiences from practical water footprint pilots and working group reports – the WFN prepared a draft of this edition. The Scientific Peer Review Committee of the Water Footprint Network reviewed the draft version of this edition and made specific recommendations with respect to revisions of the draft. The manual as it lies here is the result of incorporating the recommendations.

Also this edition will require revision in due time. All over the world research in this area is rapidly developing and more and more pilot studies on water footprint assessment are initiated, across all sectors of economy and covering all continents. In order to learn from the various ongoing practical water footprint pilot projects and from expected new scientific publications, the WFN invites both partners and non-partners to provide feedback on this edition of the manual. In this way we hope to make best use of the diverse experiences that individuals and organizations have when evaluating water footprints within different contexts and for different purposes. We aim to further refine the water footprint methodology so that it best serves the various purposes that different sectors in society see for it, at the same time striving for coherence, consistency and scientific scrutiny.

Joop de Schutter
Chair of the Supervisory Council of
the Water Footprint Network

Acronyms

CBD	Convention on Biological Diversity
CWR	crop water requirements
EPA	Environmental Protection Agency
FAO	Food and Agriculture Organization (UN)
GHG	greenhouse gas
GIEWS	Global Information and Early Warning System
GIS	geographic information system
GMIA	Global Map of Irrigation Areas
IPCC	Intergovernmental Panel on Climate Change
IRBM	integrated river basin management
IWRM	integrated water resource management
LCA	life cycle assessment
MFA	material flow analysis
MPA	maximum permissible addition
MPC	maximum permissible concentration
TMDL	total maximum daily load
UNCTAD	United Nations Conference on Trade and Development
UNDP	United Nations Development Programme
UNEP	United Nations Environment Programme
WCED	World Commission on Environment and Development
WFN	Water Footprint Network

Chapter 1

Introduction

1.1 Background

Human activities consume and pollute a lot of water. At a global scale, most of the water use occurs in agricultural production, but there are also substantial water volumes consumed and polluted in the industrial and domestic sectors (WWAP, 2009). Water consumption and pollution can be associated with specific activities, such as irrigation, bathing, washing, cleaning, cooling and processing. Total water consumption and pollution are generally regarded as the sum of a multitude of independent water demanding and polluting activities. There has been little attention paid to the fact that, in the end, total water consumption and pollution relate to what and how much communities consume and to the structure of the global economy that supplies the various consumer goods and services. Until the recent past, there have been few thoughts in the science and practice of water management about water consumption and pollution along whole production and supply chains. As a result, there is little awareness regarding the fact that the organization and characteristics of a production and supply chain strongly influence the volumes (and temporal and spatial distribution) of water consumption and pollution that can be associated with a final consumer product. Hoekstra and Chapagain (2008) have shown that visualizing the hidden water use behind products can help in understanding the global character of fresh water and in quantifying the effects of consumption and trade on water resources use. The improved understanding can form a basis for a better management of the globe's freshwater resources.

Freshwater is increasingly becoming a global resource, driven by growing international trade in water-intensive commodities. Apart from regional markets, there are also global markets for water-intensive goods such as crop and livestock products, natural fibres and bio-energy. As a result, use of water resources has become spatially disconnected from the consumers. This can be illustrated for the case of cotton. From field to end product, cotton passes through a number of distinct production stages with different impacts on water resources. These stages of production are often located in different places and final consumption can be in yet another place. For example, Malaysia does not grow cotton, but imports raw cotton from China, India and Pakistan for processing in the textile industry and exports cotton clothes to the European market (Chapagain et al, 2006b). As a result, the impacts of consumption of a final cotton product on the globe's water resources can only be found by looking at the supply chain and tracing the origins of the product. Uncovering the hidden link between consumption and water use can form the basis for the formulation of new strategies of water governance, because new triggers for change can be identified. Where final consumers, retailers, food industries and traders in water-intensive products have traditionally been out of the scope of those who studied or were responsible for good water governance, these players enter the picture now as potential 'change agents'. They can be addressed now not only in their role as *direct* water users, but also in their role as *indirect* water users.

1.2 The water footprint concept

The idea of considering water use along supply chains has gained interest after the introduction of the 'water footprint' concept by Hoekstra in 2002 (Hoekstra, 2003). The water footprint is an indicator of freshwater use that looks not only at direct water use of a consumer or producer, but also at the indirect water use. The water footprint can be regarded as a comprehensive indicator of freshwater resources appropriation, next to the traditional and restricted measure of water withdrawal. The water footprint of a product is the volume of freshwater used to produce the product, measured over the full supply chain. It is a multidimensional indicator, showing water consumption volumes by source and polluted volumes by type of pollution; all components of a total water footprint are specified geographically and temporally. The blue water footprint refers to consumption of blue water resources (surface and groundwater) along the supply chain of a product. 'Consumption' refers to loss of water from the available ground-surface water body in a catchment area. Losses occur when water evaporates, returns to another catchment area or the sea or is incorporated into a product. The green water footprint refers to consumption of green water resources (rainwater insofar as it does not become run-off). The grey water footprint refers to pollution and is defined as the volume of freshwater that is required to assimilate the load of pollutants given natural background concentrations and existing ambient water quality standards.

As an indicator of 'water use', the water footprint differs from the classical measure of 'water withdrawal' in three respects (Figure 1.1):

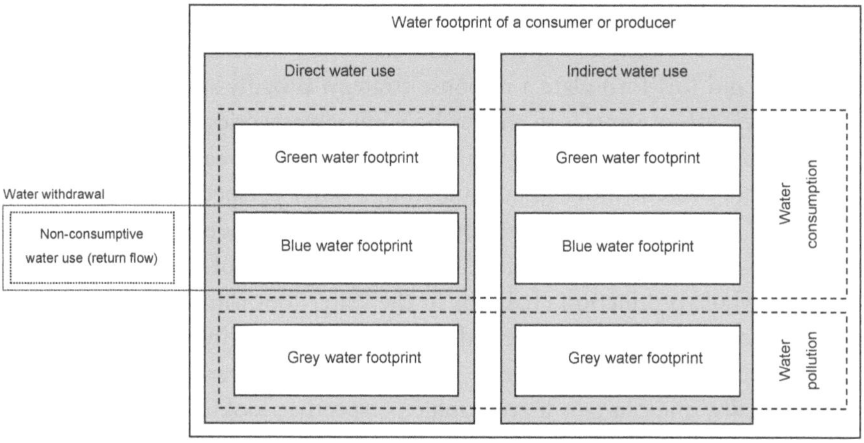

Figure 1.1 Schematic representation of the components of a water footprint. It shows that the non-consumptive part of water withdrawals (the return flow) is not part of the water footprint. It also shows that, contrary to the measure of 'water withdrawal', the 'water footprint' includes green and grey water and the indirect water-use component

1. It does not include blue water use insofar as this water is returned to where it came from.
2. It is not restricted to blue water use, but also includes green and grey water.
3. It is not restricted to direct water use, but also includes indirect water use.

The water footprint thus offers a better and wider perspective on how a consumer or producer relates to the use of freshwater systems. It is a volumetric measure of water consumption and pollution. It is *not* a measure of the severity of the local environmental impact of water consumption and pollution. The local environmental impact of a certain amount of water consumption and pollution depends on the vulnerability of the local water system and the number of water consumers and polluters that make use of the same system. Water footprint accounts give spatiotemporally explicit information regarding how water is appropriated for various human purposes. They can feed the discussion about sustainable and equitable water use and allocation and also form a good basis for a local assessment of environmental, social and economic impacts.

1.3 Water footprint assessment

'Water footprint assessment' refers to the full range of activities to: (i) quantify and locate the water footprint of a process, product, producer or consumer or to quantify in space and time the water footprint in a specified geographic area; (ii) assess the environmental, social and economic sustainability of this water footprint; and (iii) formulate a response strategy. Broadly speaking, the goal of assessing water footprints is to analyse how human activities or specific products relate to issues of water scarcity and pollution, and to see how activities and products can become more sustainable from a water perspective.

How a water footprint assessment will look, largely depends on the focus of interest. One can be interested in the water footprint of one specific process step in a whole production chain, or in the water footprint of a final product. Alternatively, one can be interested in the water footprint of a consumer or group of consumers or in the water footprint of a producer or whole economic sector. Finally, one can take a geographic perspective, looking at the total water footprint within a delineated area such as a municipality, province, nation, catchment or river basin. Such a total water footprint is the aggregation of the water footprints of many separate processes taking place in the area.

Water footprint assessment is an analytical tool, it can be instrumental in helping to understand how activities and products relate to water scarcity and pollution and related impacts and what can be done to make sure activities and products do not contribute to unsustainable use of freshwater. As a tool, a water footprint assessment provides insight, it does not tell people 'what to do'. Rather it helps people to understand what can be done.

A full water footprint assessment consists of four distinct phases (Figure 1.2):

1. Setting goals and scope.
2. Water footprint accounting.
3. Water footprint sustainability assessment.
4. Water footprint response formulation.

In order to be transparent about the choices made when undertaking a water footprint assessment study, one will have to start by clearly setting the goals and scope of the study. A water footprint study can be undertaken for many different reasons. For example, a national government may be interested in knowing its dependency on foreign water resources or it may be interested to

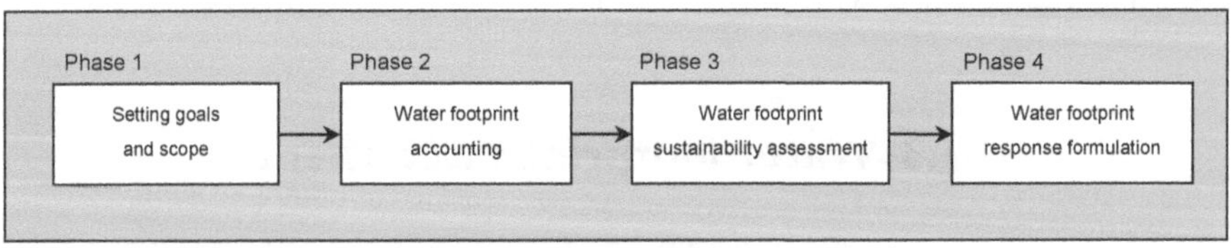

Figure 1.2 Four distinct phases in water footprint assessment

know the sustainability of water use in the areas where water-intensive import products originate. A river basin authority may be interested to know whether the aggregated water footprint of human activities within the basin violates environmental flow requirements or water quality standards at any time. The river basin authority may also want to know to what extent scarce water resources in the basin are allocated to low-value export crops. A company may be interested to know its dependence on scarce water resources in its supply chain or how it can contribute to lower the impacts on water systems throughout its supply chain and within its own operations.

The phase of water footprint accounting is the phase in which data are collected and accounts are developed. The scope and level of detail in the accounting depends on the decisions made in the previous phase. After the accounting phase is the phase of sustainability assessment, in which the water footprint is evaluated from an environmental perspective, as well as from a social and economic perspective. In the final phase, response options, strategies or policies are formulated. It is not necessary to include all the steps in one study. In the first phase of setting goals and scope, one can decide to focus only on accounting or stop after the phase of sustainability assessment, leaving the discussion about response for later. Besides, in practice, this model of four subsequent phases is more a guideline than a strict directive. Returning to earlier steps and iteration of phases may often be necessary. In first instance, a company may be interested in a rough exploration of all phases, in order to identify critical components in its water footprint and set priorities for response, while later on it may like to seek much greater detail in certain areas of the accounts and the sustainability assessment.

1.4 Guide for the reader

The four phases of water footprint assessment are addressed in the following chapters. Chapter 2 shows the important issues that have to be considered when setting the goals and scope in water footprint assessment. Chapter 3 contains the definitions and methods for water footprint accounting. Chapter 4 gives guidelines for the stage of water footprint sustainability assessment. Chapter 5 gives an overview of water footprint response options to be considered in the stage of policy formulation. Chapter 6 puts the method of water footprint assessment in a wider context and discusses its limitations. Chapter 7 identifies and discusses the major challenges to be addressed in the future. Chapter 8 is the concluding chapter. Depending on the interest of the reader, one can focus on different parts of the manual. Particularly in Chapter 3, on water footprint accounting, the reader can be selective, depending on whether the reader takes the perspective of a consumer (focus on Section 3.5), national government (Section 3.7), river basin authority (Section 3.8) or businesses (Section 3.10). One will see that the basics of water footprint accounting – process and product accounts (Sections 3.3 and 3.4) – are relevant for all water footprint applications.

In the course of the text, various concepts will be defined. In order to make it easy for the reader to look up the definitions of key terms used in this manual, a glossary is included in the back of the book. Another helpful section in the book is Appendix VI, which addresses the most frequently asked questions in the context of water footprint assessment.

Chapter 2

Goals and Scope of Water Footprint Assessment

2.1 Goals of water footprint assessment

Water footprint studies may have various purposes and be applied in different contexts. Each purpose requires its own scope of analysis and will allow for different choices when making assumptions. One can assess the water footprint of different entities, so it is most important to start specifying in which water footprint one is interested. One can be interested, for instance, in the:

- water footprint of a process step
- water footprint of a product
- water footprint of a consumer
- water footprint of a group of consumers
  - water footprint of consumers in a nation
  - water footprint of consumers in a municipality, province or other administrative unit
  - water footprint of consumers in a catchment area or river basin
- water footprint within a geographically delineated area
  - water footprint within a nation
  - water footprint within a municipality, province or other administrative unit
  - water footprint within a catchment area or river basin
- water footprint of a business
- water footprint of a business sector
- water footprint of humanity as a whole

A checklist for defining the goal of water footprint assessment is given in Box 2.1. The list is not exhaustive but rather shows a number of things to be specified. Probably the most important question is what sort of detail one seeks. If the purpose is awareness-raising, national or global average estimates for the water footprints of products are probably sufficient. When the goal is hotspot identification, one will need to include a greater detail in the scope and subsequent accounting and assessment, so that it is possible to pinpoint exactly where and when the water footprint has the greatest local environmental, social or economic impacts. If the aim is to formulate policy and establish targets on quantitative water footprint reduction, an even higher degree of spatial and temporal detail is required. Besides, one will have to embed the water footprint assessment in a broader deliberation incorporating factors other than water alone (see also Chapter 6).

Box 2.1 *Goals of water footprint assessment*

General

- What is the ultimate target? Awareness-raising, hotspot identification, policy formulation or quantitative target setting?
- Is there a focus on one particular phase? Focus on accounting, sustainability assessment or response formulation?
- What is the scope of interest? Direct and/or indirect water footprint? Green, blue and/or grey water footprint?
- How to deal with time? Aiming at assessment for one particular year or at the average over a few years, or trend analysis?

Process water footprint assessment

- What process to consider? One specific process or alternative, substitutable processes (in order to compare the water footprints of alternative techniques)?
- What scale? One specific process in a specific location or the same process in different locations?

Product water footprint assessment

- What product to consider? One stock-keeping unit of a particular brand, one particular sort of product or a whole product category?
- What scale? Include product(s) from one field or factory, one or more companies or one or more production regions?

Consumer or community water footprint assessment

- Which community? One individual consumer or the consumers within a municipality, province or state?

Assessment of the water footprint within a geographically delineated area

- What are the area boundaries? A catchment, river basin, municipality, province, state or nation?

- What is the field of interest? Examine how the water footprint within the area is reduced by importing virtual water and how the water footprint within the area is increased by making products for export, analyse how the area's water resources are allocated over various purposes, and/or examine where the water footprint within the area violates local environmental flow requirements and ambient water quality standards?

National water footprint assessment (water footprint within a nation and water footprint of national consumption)

- What is the scope of interest? Assess the water footprint within a nation and/or the water footprint of national consumption? Analyse the internal and/or the external water footprint of national consumption?
- What is the field of interest? Assess national water scarcity, sustainability of national production, export of scarce water resources in virtual form, national water saving by import of water in virtual form, sustainability of national consumption, impacts of the water footprint of national consumption in other countries and/or dependency on foreign water resources?

Business water footprint assessment

- What is the scale of study? A company unit, whole company or a whole sector? (When the scale of interest is the product level, see above under product water footprint assessment.)
- What is the scope of interest? Assess the operational and/or the supply chain water footprint?
- What is the field of interest? Business risk, product transparency, corporate environmental reporting, product labelling, benchmarking, business certification, identification of critical water footprint components, formulation of quantitative reduction targets?

2.2 Scope of water footprint accounting

One will have to be clear and explicit about the 'inventory boundaries' when setting up a water footprint account. The inventory boundaries refer to 'what to include' and 'what to exclude' from the accounts and should be chosen as a function of the purpose of the account. One can use at least the following checklist when setting up a water footprint account:

- Consider blue, green and/or grey water footprint?
- Where to truncate the analysis when going back along the supply chain?
- Which level of spatiotemporal explication?
- Which period of data?

- For consumers and businesses: consider direct and/or indirect water footprint?
- For nations: consider water footprint within the nation and/or water footprint of national consumption; consider internal and/or external water footprint of national consumption?

Blue, green and/or grey water footprint?
Blue water resources are generally scarcer and have higher opportunity costs than green water, so that may be a reason to focus on accounting the blue water footprint only. However, green water resources are also limited and thus scarce, which gives an argument to account the green water footprint as well. Besides, green water can be substituted by blue water – and in agriculture the other way around as well – so that a complete picture can be obtained only by accounting for both. The argument for including green water use is that the historical engineering focus on blue water has led to the undervaluation of green water as an important factor of production (Falkenmark, 2003; Rockström, 2001). The idea of the grey water footprint was introduced in order to express water pollution in terms of a volume polluted, so that it can be compared with water consumption, which is also expressed as a volume (Chapagain et al, 2006b; Hoekstra and Chapagain, 2008). If one is interested in water pollution and in comparing the relative claims of water pollution and water consumption on the available water resources, it is relevant to account the grey in addition to the blue water footprint.

Where to truncate the analysis when going back along the supply chain?
The truncation issue is a basic question in water footprint accounting. One faces similar questions as in carbon and ecological footprint accounting, energy analysis and life cycle assessment. No general guidelines have been developed yet in the field of water footprint accounting, but the general rule is: include the water footprint of all processes within a production system (production tree) that 'significantly' contribute to the overall water footprint. The question remains what 'significant' is; one can say for instance 'larger than 1 per cent' (or 'larger than 10 per cent' when interested in the largest components only). If one traces the origins of a particular product, one will see that supply chains are never-ending and widely diverging because of the variety of inputs used in each process step. In practice, however, there are only a few process steps that substantially contribute to the total water footprint of the final product. As a rule of thumb, one can expect that, when a product includes ingredients that originate from agriculture, those ingredients often give a major contribution to the overall water footprint of the product. This is the case because an estimated 86 per cent of the water footprint of humanity is within the agricultural sector

(Hoekstra and Chapagain, 2008). Industrial ingredients are likely to contribute particularly when they can be associated with water pollution (so they will contribute to the grey water footprint).

A specific question that falls under the truncation issue is whether one should account for the water footprint of labour, which is an input factor in nearly all processes. The argument could be made that employees are an input factor that requires food, clothing and drinking water, so that all the direct and indirect water requirements of employees should be included in the indirect water footprint of a product. However, this creates a very serious accounting problem, well-known in the field of life cycle assessment. The problem is that double counting would occur. The underlying idea of natural resources accounting of products is to allocate all natural resource use to the final consumer products and based on consumption data to consumers. All natural resource use is thus ultimately attributed to consumers. Consumers are, however, also workers. It would create a never-ending loop of double, triple counting and so on, when the natural resource use attributed to a consumer would be counted as natural resource use underlying the input factor labour in production. In short, it is common practice to exclude labour as a factor embodying indirect resource use.

Another specific question often posed – particularly by analysts who have experience with carbon footprint accounting – is whether the water footprint of transport should be included. Transport consumes a lot of energy, the amount of which may constitute a significant component of the overall energy used to produce a product and get it to its final destination. In many cases, transport does not consume a significant amount of freshwater if compared to the total freshwater consumed to make and transport a product. It depends on the type of product and the type of energy applied. In general, whether the water footprint of transport is to be included in the analysis depends on the rule chosen with respect to how to truncate the analysis. When transport is expected to have a minor contribution to the overall water footprint of a product, the component can be left out of the analysis. We particularly recommend including the water footprint of transport when biofuels or hydropower are used as the source of energy, because these forms of energy are known to have a relatively large water footprint per unit of energy. More in general, one can ask whether the water footprint of energy applied in a production system should be included in the assessment of the water footprint of the final product. Again, in most cases the contribution of the factor energy will be a small percentage of the overall water footprint of a product. An exception may be when energy is sourced from biofuel or from electricity from biomass combustion or hydropower, because those forms of energy have a relatively large water footprint per unit of energy (Gerbens-Leenes et al, 2009a, b; Yang et al, 2009; Dominguez-Faus et al, 2009).

Table 2.1 *Spatiotemporal explication in water footprint accounting*

	Spatial explication	*Temporal explication*	*Source of required data on water use*	*Typical use of the accounts*
Level A	Global average	Annual	Available literature and databases on typical water consumption and pollution by product or process.	Awareness-raising; rough identification of components contributing most to the overall water footprint; development of global projections of water consumption.
Level B	National, regional or catchment-specific	Annual or monthly	As above, but use of nationally, regionally or catchment specific data.	Rough identification of spatial spreading and variability; knowledge base for hotspot identification and water allocation decisions.
Level C	Small catchment or field-specific	Monthly or daily	Empirical data or (if not directly measurable) best estimates on water consumption and pollution, specified by location and over the year.	Knowledge base for carrying out a water footprint sustainability assessment; formulation of a strategy to reduce water footprints and associated local impacts.

Note: The three levels can be distinguished for all forms of water footprint accounting (for example, product, national, corporate accounts).

Which level of spatiotemporal explication?

Water footprints can be assessed at different levels of spatiotemporal detail (Table 2.1). At Level A, the lowest level of detail, the water footprint is assessed based on global average water footprint data from an available database. The data refer to multi-year averages. This level of detail is sufficient and even most instrumental for the purpose of awareness-raising. This level of detail can also be suitable when the aim is to identify products and ingredients that most significantly contribute to the overall water footprint. Global-average water footprint data can also be useful for developing rough projections of future global water consumption given major changes in consumption patterns (such as a shift towards more meat or bio-energy). At Level B, the water footprint is assessed based on national or regional average or catchment-specific water footprint data from an available geographically explicit database. The water footprints are preferably specified by month, but will still be multi-year average monthly data. This level of accounting is suitable to provide a basis for understanding where hotspots in local watersheds can be expected and for making water allocation decisions. At Level C, water footprint accounts are geographically and temporally explicit, based on precise data on inputs used, and precise sources of those inputs. The minimum spatial resolution is the level of small catchments (~100–1000 km^2), but if desired and when data allow, one can account at field level. In the latter case, we are talking about accounts that map water footprint per farm, living district or industry. The minimum temporal resolution is a month and studying inter-annual variations will be part of the analysis. The accounting is based on best estimates of actual local water consumption and pollution, preferably verified on the ground. This high level of spatiotemporal detail is suitable for formulating site-specific water footprint reduction strategies.

Which period of data?

Water availability fluctuates within a year and across years as well. As a consequence of varying water availability, water demand varies in time as well. One should therefore be extremely cautious in making claims regarding a water footprint trend in time. Whatever water footprint study is undertaken, one should be explicit regarding the period of data used, because the period chosen will affect the outcome. In dry years, the blue water footprint of a crop product will be much higher than in wet years, because more irrigation water will be required. One can choose to calculate water footprints for one particular year or a number of specific years, but alternatively one can choose to calculate the water footprint under an average year given the existing climate (defined as the average over a consecutive period of 30 years). In the latter case, one will combine different periods in one analysis: one takes, for example, production and yield data for a recent period of five years but data on climate (temperature and precipitation) as an average for the past 30 years.

Direct and/or indirect water footprint?

The general recommendation is to include both direct and indirect water footprints. Whereas direct water footprints are the traditional focus of consumers and companies, the indirect water footprint is generally much larger. By addressing only their direct water footprint, consumers would neglect the fact that the largest part of their water footprint is associated with the products they buy in the supermarket or elsewhere, not the water they consume at home. For most businesses, the water footprint in their supply chain is much bigger than the water footprint of their own operations; ignoring the supply chain component may lead to investments in making improvements in the operational water use while investments in improving the supply chain could have been more cost-effective. Depending on the purpose of a particular study, however, one can of course decide to include only the direct or indirect water footprint in the analysis. There is some similarity here with the 'scopes' as distinguished in carbon footprint accounting (see Box 2.2).

Box 2.2 *Are there 'scopes' in water footprint accounting as there are in the case of corporate carbon footprint accounting?*

A carbon footprint is the total set of greenhouse gas (GHG) emissions caused directly and indirectly by an individual, organization, event or product. In the field of corporate carbon footprint accounting, three 'scopes' have been defined (WRI and WBCSD, 2004). Scope 1 refers to the accounting of 'direct' GHG emissions, which occur from sources owned or controlled by the company. Examples are: emissions from combustion in owned or controlled boilers, furnaces, vehicles and so on; emissions from chemical production in owned or controlled process equipment. Scope 2 refers to accounting of 'indirect' GHG emissions from the generation of purchased electricity consumed by the company. Scope 3 refers to other indirect GHG emissions, which are a consequence of the activities of the company but occur from sources not owned or controlled by the company. Examples of Scope 3 activities are: extraction and production of purchased materials; transportation of purchased fuels; and use of sold products and services. The distinction between 'direct' and 'indirect' is also made in case of water footprint accounting. The total water footprint of a consumer or producer refers, by definition, to both the direct and the indirect water use of this consumer or producer. This means that, without specification, the term water footprint refers to the sum of direct and indirect. The distinction between Scopes 2 and 3 as applied in carbon footprint accounting is not useful in the case of water footprint accounting. In water footprint accounting there are thus two 'scopes' only: 'direct' and 'indirect' water footprint.

Consider the water footprint within a nation or the water footprint of national consumption?

The 'water footprint within a nation' refers to the total freshwater volume consumed or polluted within the territory of the nation. This includes water use for making products consumed domestically but also water use for making export products. The 'water footprint within a nation' is different from the 'water footprint of national consumption', which refers to the total amount of water used to produce the goods and services consumed by the inhabitants of the nation. This refers to both water use within the nation and water use outside the territory of the nation, but is restricted to the water use behind the products consumed within the nation. The water footprint of national consumption thus includes an internal and an external component. Including an analysis of the external water footprint is key in order to get a complete picture of how national consumption translates to water use not only in the country itself but also abroad, and thus to analyse water dependency and sustainability of imports.

Looking at the water footprint within a nation is sufficient when the interest lies with the use of domestic water resources only.

2.3 Scope of water footprint sustainability assessment

For the phase of sustainability assessment, the primary question is whether one takes a geographic perspective or a process, product, consumer or producer perspective. In the case of a geographic perspective, one will look at the sustainability of the aggregated water footprint in a certain area, preferably a catchment area or a whole river basin, because this is the natural unit in which one can easily compare water footprint and water availability and where allocation of water resources and potential conflicts take place. In the case of a process, product, consumer or producer perspective, the focus is not on the *aggregate* water footprint in one geographic setting, but on the *contribution* of the water footprint of the individual process, product, consumer or producer to the larger picture. The question of the contribution contains two elements: (i) what is the contribution of the specific process, product, consumer or producer water footprint to the global water footprint of humanity, and (ii) what is its contribution to the aggregated water footprints in specific geographic areas? The contribution to the global total is interesting from a sustainability point of view, because the world's freshwater resources are limited, so there should be concern with any contribution beyond the reasonable maximum need from a technical or societal point of view. The contribution to the aggregated water footprints in specific catchments or river basins is interesting because there should be concern with any contribution that takes place in a catchment or river basin where the water footprint results in a situation where basic environmental needs are not fulfilled or where water allocation is socially or economically unsustainable.

The scope of a water footprint sustainability assessment thus primarily depends on the perspective chosen. In all cases, the scope needs to be further specified depending on the goals of the assessment. In the case of a geographic perspective, one can use the following checklist:

- Consider the sustainability of the green, blue and/or grey water footprint?
- Consider the environmental, social and/or economic dimension of sustainability?
- Identify hotspots only or analyse in detail primary and/or secondary impacts in the hotspots as well?

The answer to the last point will influence the required level of detail in the assessment. Identifying hotspots – in other words, finding the (sub)catchments in which the water footprint is unsustainable during specific times of the year – can be done by comparing green and blue water footprints to green and blue water availability and by comparing grey water footprints to available assimilation capacity, without the need for analysing in any detail the primary and secondary impacts that may occur as a result of water scarcity or pollution. The finer the spatial and temporal resolution level applied when comparing water footprints to water availability, the better one can localize the hotspots. Looking at annual values at the level of whole river basins, results in a very crude localization of hotspots. When the goal is to achieve more accuracy, it is necessary to look at monthly values and at the level of smaller catchments. When the goal goes beyond hotspot identification and includes a better understanding of what a water footprint in a geographic area really implies, one needs to describe in detail how the water footprint within a catchment affects water flows and quality in the area (primary impacts) and how this finally impacts upon ultimate indicators like welfare, social equity, human health and biodiversity.

When the interest is in the sustainability of the water footprint of a process, product, consumer or producer, the focus will be on exploring (i) whether the water footprint unnecessarily contributes to the global water footprint of humanity and (ii) whether the water footprint contributes to specific hotspots. For the purpose of the former, one can suffice with comparing each separate process or product water footprint with a global benchmark for that process or product, when such a benchmark already exists. In the absence of such benchmarks, the scope of the assessment will need to be extended so that it also includes studying what could be a reasonable benchmark. For exploring whether the water footprint of a process, product, consumer or producer contributes to specific hotspots, one could suffice checking for each water footprint component whether it is located in a hotspot or not. This requires a worldwide hotspot database at the level of the spatial and temporal detail demanded. When such background data are not available, the scope of study has to be extended in order to include catchment studies from the geographic perspective as well, for all the catchments where the (major) components of the water footprint of the process, product, consumer or producer are located.

2.4 Scope of water footprint response formulation

The scope of the response formulation phase depends again on the sort of water footprint one is looking at. In the case of the water footprint within a geographically delineated area, the question is: what can be done by who to reduce the water footprint within that area, by how much and what time path? When setting the scope for response formulation, one will have to be particularly clear about 'response by who'. One can look at what governments can do – which is what people will probably think of first when talking about the water footprint within a geographic setting – but one can also look at what, for example, consumers, farmers, companies and investors can do and what may have to be done through intergovernmental cooperation. And with respect to the government, one can distinguish between different levels of government and different governmental bodies at each level. At the national level, for example, required response may translate to actions within different ministries, ranging from the ministries of water, the environment, agriculture, energy and spatial planning to the ministries of economics, trade and foreign affairs. When setting the scope for identifying response measures, it is important to be clear from the beginning the angle(s) from which one will identify those measures.

In the case of the water footprint of a consumer or community of consumers, one can simply look at what the consumer(s) can do, but here also one can include an analysis of what others – in this case for instance companies and governments – can do. When considering response in the context of assessing a company's water footprint, it is most logical to look, at least, at what sort of response the company can develop itself, but here also the scope can be formulated broader.

Chapter 3

Water Footprint Accounting

3.1 Human appropriation of fresh water: What do we measure and why?

Water on Earth is constantly moving. Water evaporates from the soil and open water surfaces as a result of solar and wind energy. In addition, plants draw water up from the soil and release it to the atmosphere through the stomata on their leaves, a process called transpiration. The processes of evaporation and transpiration together are called 'evapotranspiration' (although in daily language, the short term evaporation is generally used to include transpiration as well). The amount of water in the atmosphere increases through evapotranspiration, but decreases again through precipitation. Within the atmosphere, water vapour moves around the globe according to complex patterns, so water that evaporates in one place does not necessarily return as precipitation in the same place. The amount of water on land increases as a result of precipitation but decreases as a result of evapotranspiration. Since precipitation on land exceeds evapotranspiration (not on a daily scale, but in the long run), there is a water surplus on land, which leads to run-off. Run-off water from land finally ends up in the ocean. While lands have a precipitation surplus, oceans have an evaporation surplus. Overall, there is net transport of water from oceans to land through the atmosphere. It is brought back from land to ocean through run-off. Run-off occurs partly through overland flow (rivers and streams) and partly through groundwater flow. The volume of water on Earth remains more or less equal.

For nearly all human purposes, we need freshwater as it occurs on land. Salt water as it occurs in the ocean is not useful for drinking, washing, cooking, field irrigation or for most applications in industry. Salt water can be desalinized, but this is a costly and energy-intensive process, feasible for a limited number of applications only. Besides, salt water is available at the coast, while much of the water needs are inland, so that transport uphill becomes an issue as well. In short, humans mainly depend on freshwater as it occurs on land. Although water forms a cycle, so that freshwater on land is continuously replenished, its availability is not unlimited. Per year, people need a certain volume of water for domestic, agricultural and industrial purposes, which cannot exceed the annual replenishment rate. The major question is therefore: how much freshwater is available over a certain period and what is man's actual appropriation of this flow in this period? Water footprint accounting provides the data for answering the second half of the question. The water footprint basically expresses human appropriation of freshwater in volume terms. Comparing man's water footprint with the actual freshwater availability is part of a water footprint sustainability assessment, which is the subject of Chapter 4.

For understanding freshwater appropriation by humans in relation to the hydrological cycle, one can consider a river basin. A river basin is the entire geographical area drained by a river and its tributaries. All run-off from a river basin is conveyed to the same outlet. Other terms for 'river basin' often used are 'catchment area', 'drainage basin', 'drainage area' and 'watershed'. The total annual water availability in a catchment area is given by the annual volume of precipitation. When we ignore possible, but generally small, changes in water storage in a catchment area, the total annual precipitation flow will leave the basin again, partly through evapotranspiration and partly through run-off from the catchment. Both the evaporative flow and the run-off can be appropriated by humans. The green water footprint refers to the human use of the evaporative flow from the land surface, mostly for growing crops or production forest (Figure 3.1). The blue water footprint refers to the consumptive use of the run-off flow, in other words, the abstraction of run-off from the catchment insofar as it does not return to the catchment in the form of return flow.

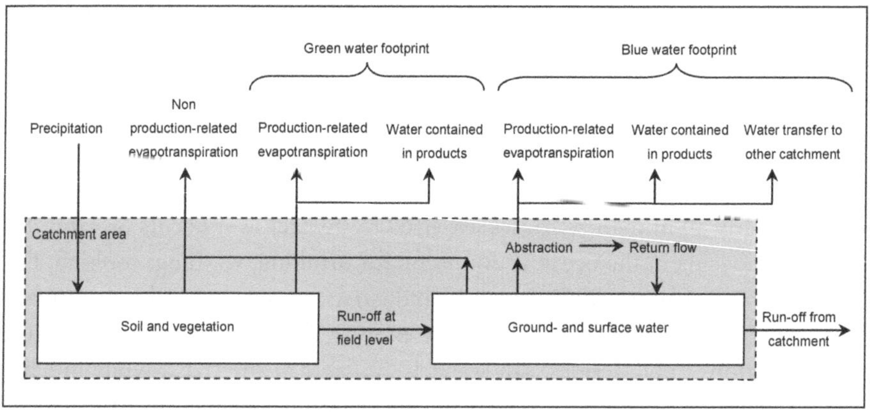

Figure 3.1 The green and blue water footprint in relation to the water balance of a catchment area

Historically, people have used run-off flows both as a source of fresh water and as a drain for their waste. Obviously, there are limits to using run-off flows as a source or sink. The total run-off flow has a limited volume to tap from and a limited capacity to assimilate waste. The blue water footprint shows the volume that has been effectively taken out of the total run-off flow, so it shows the 'appropriated tap capacity'. The grey water footprint shows the 'appropriated waste assimilation capacity'. It is defined as the volume of water required to assimilate waste, quantified as the volume of water needed to dilute pollutants to such an extent that the quality of the ambient water remains above agreed water quality standards. The advantage of measuring water pollution in terms of water volumes appropriated is that different forms of pollution are brought into one denominator, namely the water volume appropriated for waste assimilation. Besides, when water pollution is expressed in the same terms as water consumption, one can compare the use of run-off as a source (the blue water footprint) to the use of run-off as a sink (the grey water footprint).

3.2 Coherence between different sorts of water footprint accounts

The water footprint of one single 'process step' is the basic building block of all water footprint accounts (see Figure 3.2 and Box 3.1). The water footprint of an intermediate or final 'product' (good or service) is the aggregate of the water footprints of the various process steps relevant in the production of the product. The water footprint of an individual consumer is a function of the water footprints of the various products consumed by the consumer. The water footprint of a community of consumers – for example, the inhabitants of a municipality, province, state or nation – is equal to the sum of the individual water footprints of the members of the community. The water footprint of a producer or whatever sort of business is equal to the sum of the water footprints of the products that the producer or business delivers. The water footprint within a geographically delineated area – be it a province, nation, catchment area or river basin – is equal to the sum of the water footprints of all processes taking place in that area. The total water footprint of humanity is equal to the sum of the water footprints of all consumers of the world, which is equal to the sum of the water footprints of all final consumer goods and services consumed annually and also equal to the sum of all water-consuming or polluting processes in the world.

Water footprints of final (consumer) products can be added without double counting. This is due to the fact that process water footprints are always

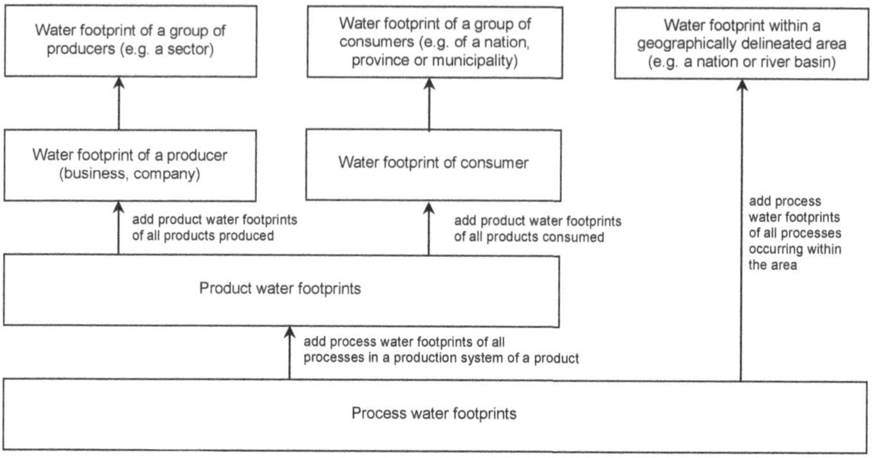

Figure 3.2 Process water footprints as the basic building block for all other water footprints

exclusively allocated to one final product or, when a process contributes to more than one final product, a process water footprint is divided over the different final products. Adding water footprints of intermediate products does not make sense, because double counting can easily occur. If one would add, for instance, the water footprint of cotton fabric and the water footprint of harvested cotton, one would double count, because the former includes the latter. Similarly, one can add the water footprints of individual consumers without double counting, but one should not add the water footprints of different producers as this can lead to double counting.

The water footprint of consumers is related to the water footprints of the producers in the supply-chain. Figure 3.3 shows a simplified example of the supply chain of an animal product. The total water footprint of a consumer is the sum of its direct and indirect water footprints. When we focus on meat consumption, the direct water footprint of the consumer refers to the volume of water consumed or polluted when preparing and cooking the meat. The indirect water footprint of the meat consumer depends on the direct water footprints of the retailer that sells the meat, the food processor that prepares the meat for sale, the livestock farm that raises the animal and the crop farm that produces the feed for the animal. The indirect water footprint of the retailer depends on the direct water footprints of the food processor, livestock farm and crop farm, and so on.

The 'water footprint of the consumers in an area' is not equal to the 'water footprint within the area', but they are related. Figure 3.4 shows the relation between the water footprint of national consumption and the water footprint within a nation in a simplified example for two trading nations. The 'internal'

Box 3.1 *The relation between the different sorts of water footprints*

- The water footprint of a product = the sum of the water footprints of the process steps taken to produce the product (considering the whole production and supply chain).
- The water footprint of a consumer = the sum of the water footprints of all products consumed by the consumer.
- The water footprint of a community = the sum of the water footprints of its members.
- The water footprint of national consumption = the sum of the water footprints of its inhabitants.
- The water footprint of a business = the sum of the water footprints of the final products that the business produces.
- The water footprint within a geographically delineated area (for example, a municipality, province, state, nation, catchment or river basin) = the sum of the process water footprints of all processes taking place in the area.

water footprint of national consumption is equal to the water footprint within the nation insofar as it is not related to producing export products. The 'external' water footprint of national consumption can be found by looking at the import of products (and thus water in virtual form) and the associated water footprint within the exporting nation.

A water footprint is expressed in terms of a water volume per unit of product or as a water volume per unit of time (Box 3.2). The water footprint of a process is expressed as water volume per unit of time. When divided over the quantity of product that results from the process, it can also be expressed as water volume per product unit. A product water footprint is always expressed in terms of water volume per unit of product (usually m^3/ton or litre/kg). The water footprint of a consumer or producer or the water footprint within an area is always expressed as water volume per unit of time. Depending on the level of detail that one aims to provide, the water footprint can be expressed per day, month or year.

3.3 Water footprint of a process step

3.3.1 Blue water footprint

The blue water footprint is an indicator of consumptive use of so-called blue water, in other words, fresh surface or groundwater. The term 'consumptive water use' refers to one of the following four cases:

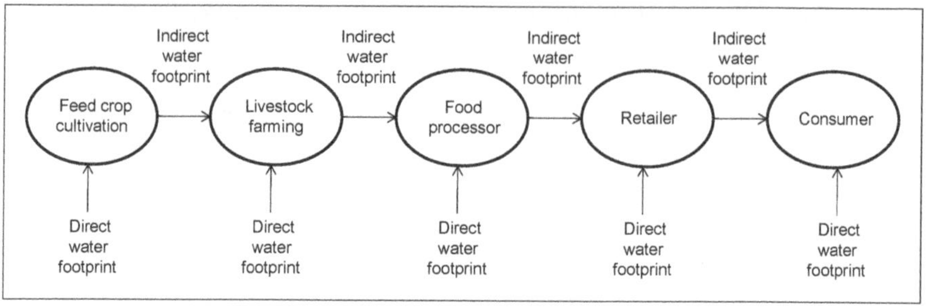

Figure 3.3 The direct and indirect water footprint in each stage of the supply chain of an animal product

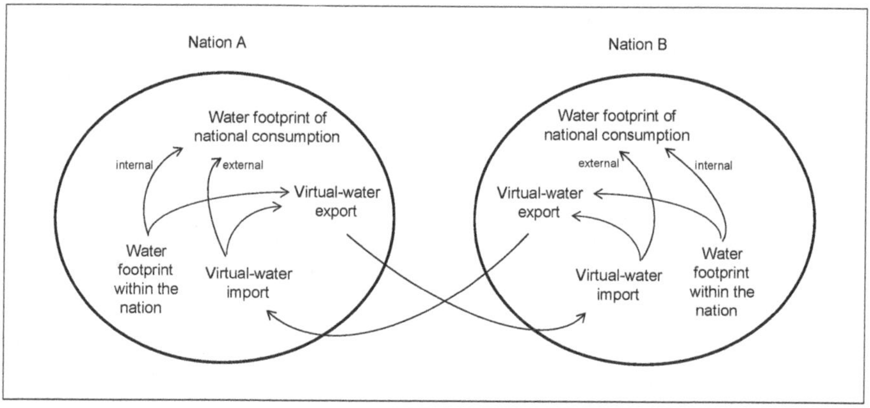

Figure 3.4 The relation between the water footprint of national consumption and the water footprint within a nation in a simplified example for two trading nations

1. Water evaporates;
2. Water is incorporated into the product;
3. Water does not return to the same catchment area, for example, it is returned to another catchment area or the sea;
4. Water does not return in the same period, for example, it is withdrawn in a scarce period and returned in a wet period.

The first component, evaporation, is generally the most significant one. Therefore one will often see that consumptive use is equated with evaporation, but the other three components should be included when relevant. All production-related evaporation counts, including the water that evaporates during water storage (for example, from artificial water reservoirs), transport (for example,

Box 3.2 *Unit of a water footprint*

- The water footprint of a process is expressed as water volume per unit of time. When divided over the quantity of product that results from the process (product units per unit of time), it can also be expressed as water volume per product unit.
- The water footprint of a product is always expressed as water volume per product unit. Examples:
  - water volume per unit of mass (for products where weight is a good indicator of quantity)
  - water volume per unit of money (for products where value tells more than weight)
  - water volume per piece (for products that are counted per piece rather than weight)
  - water volume per unit of energy (per kcal for food products, or per joule for electricity or fuels)
- The water footprint of a consumer or business is expressed as water volume per unit of time. It can be expressed as water volume per monetary unit when the water footprint per unit of time is divided by income (for consumers) or turnover (for businesses). The water footprint of a community of consumers can be expressed in terms of water volume per unit of time per capita.
- The water footprint within a geographically delineated area is expressed as water volume per unit of time. It can be expressed in terms of water volume per monetary unit when divided over the income in the area.

from open canals), processing (for example, evaporation of heated water that is not recollected) and collection and disposal (for example, from drainage canals and from wastewater treatment plants).

'Consumptive water use' does not mean that the water disappears, because water will remain within the cycle and always return somewhere. Water is a renewable resource, but that does not mean that its availability is unlimited. In a certain period, the amount of water that recharges groundwater reserves and that flows through a river is always limited to a certain amount. Water in rivers and aquifers can be used for irrigation or industrial or domestic purposes. But in a certain period one cannot consume more water than is available. The blue water footprint measures the amount of water available in a certain period that is consumed (in other words, not immediately returned within the same catchment). In this way, it provides a measure of the amount of available blue water consumed by humans. The remainder, the groundwater and surface water flows not consumed for human purposes, is left to sustain the ecosystems that depend on the groundwater and surface water flows.

The blue water footprint in a process step is calculated as:

$$WF_{proc,blue} = BlueWaterEvaporation + BlueWaterIncorporation + LostReturnflow \quad \text{[volume/time]} \qquad (1)$$

The last component refers to the part of the return flow that is not available for reuse within the same catchment within the same period of withdrawal, either because it is returned to another catchment (or discharged into the sea) or because it is returned in another period of time.

In assessing the blue water footprint of a process it may be relevant (depending on the scope of the study) to distinguish between different sorts of blue water sources. The most relevant division is between surface water, flowing (renewable) groundwater and fossil groundwater. One can make the distinction by speaking respectively of the blue surface water footprint, the blue renewable groundwater footprint and the blue fossil groundwater footprint (or the light-blue, dark-blue and black water footprint if one really likes the use of the colours). In practice, it is often very difficult to make the distinction because of insufficient data, which is the reason the distinction is often not made. It is possible, however, if data allow, to specify the blue water footprint by source (see examples in Aldaya and Llamas, 2008; Aldaya and Hoekstra, 2010; Mekonnen and Hoekstra, 2010b).

When specifying the total blue water footprint by source, one may also like to explicitly distinguish the consumptive use of harvested rainwater. Rainwater harvesting is a bit of a particular case, since one may argue whether harvested rainwater is green or blue water. Mostly, rainwater harvesting refers to the collection of rain that otherwise would become run-off. Since consumptive use of harvested rainwater will subtract from run-off, we recommend to consider such water use as a blue water footprint. Various sorts of rainwater harvesting techniques exist to provide drinking water, water for livestock or water for irrigating crops or gardens. As long as one speaks of local collection of run-off – as in the case of rainwater harvesting from rooftops or other hard surfaces or in the case of leading the rain to small ponds – one can categorize consumptive use of this water under the blue water footprint. If, on the contrary, one speaks of measures to increase the soil water holding capacity or about green rooftops to retain rainwater, consumptive use of this water for crop production will fall under the green water footprint.

The unit of the blue process water footprint is water volume per unit of time, for example, per day, month or year. When divided over the quantity of product that stems from the process, the process water footprint can also be expressed in terms of water volume per unit of product. In Box 3.3 we reflect on where to obtain the data necessary for blue water footprint accounting.

Box 3.3 *Data sources for the calculation of a blue water footprint*

Industrial processes: Each component of the blue process water footprint can be measured, directly or indirectly. It is generally known how much water is added in order to become part of the product. How much water evaporates during storage, transport, processing and disposal is generally not measured directly, but can be inferred from the difference between abstraction and final disposal volumes. Ideally, one can rely on databases that contain typical data on consumptive water use for various types of manufacturing processes. Such databases, however, do hardly exist and generally contain data on water withdrawals (abstractions), not on consumptive water use. Besides, these databases generally lack the necessary details and contain data on water use per industrial sector (for example, sugar refineries, textile mills, paper mills and so on) rather than per manufacturing process. Two data-rich compendiums are Gleick (1993) and Van der Leeden et al (1990), but both are US-focused and mainly limited to data on water withdrawals. One can also consult proprietary databases such as Ecoinvent (2010) but such databases generally provide data on water withdrawals, not consumptive water use. The best sources for blue water consumption in manufacturing processes are the manufacturers themselves or regional or global branch organizations.

Agricultural processes: Available statistics on blue water use in agriculture generally show total water withdrawals for irrigation only, not blue *consumptive* water use. Measuring water evapotranspiration from a field is a laborious task. And even when total evapotranspiration was measured, one would need to estimate which part of the total is blue water. Therefore one will generally rely on water balance models that use data on climate, soil, crop characteristics and actual irrigation as input. Section 3.3.4 shows in more detail how one can estimate the blue water footprint in crop growth based on a water balance model. Based on global maps on where different crops are grown and on global maps of climate, soils and irrigation, a few research groups in the world have started to make spatial-explicit estimates of blue (and green) water footprints of crop growing. For wheat alone, for example, four global datasets are available: Liu et al (2007, 2009), Siebert and Döll (2010), Mekonnen and Hoekstra (2010a) and Zwart et al (2010). At the website of the Water Footprint Network – www.waterfootprint.org – geographically explicit data on the water footprint of crop growing are available for all major crops in the world. These datasets can be used for water footprint accounting at Level B (see Table 2.1). For accounting at Level C, one will have to apply an appropriate water balance model oneself, together with locally specific input data.

We finalize this section by considering two specific cases for which it may not be immediately clear to the reader how to properly account. The first case concerns the issue of water recycling and reuse. The second case concerns the question of how to account in the case of an inter-basin water transfer.

Water recycling and reuse

Water recycling and reuse are often used as two interchangeable terms. Here we define 'water recycling' specifically as the *on-site* reuse of water *for the same purpose* and 'water reuse' as the reuse of water elsewhere, possibly for another purpose. In the case of recycling, we can make an additional distinction between recycling of wastewater (by treating it for reuse) and recycling of evaporated water (by condensing water vapour for reuse). The different sorts of water recycling and reuse are shown for a simple example in Figure 3.5. The figure shows two processes, the second of which reuses (treated) wastewater from the first. The scheme shows that for the blue water footprints of both processes it is the consumptive water use (evaporation and incorporation into products) that counts. Water recycling and reuse can be instrumental in reducing the blue water footprint of a process only when it effectively reduces consumptive water use. Water recycling and reuse may also be instrumental in reducing the grey water footprint of water users, but this will be discussed in Section 3.3.3.

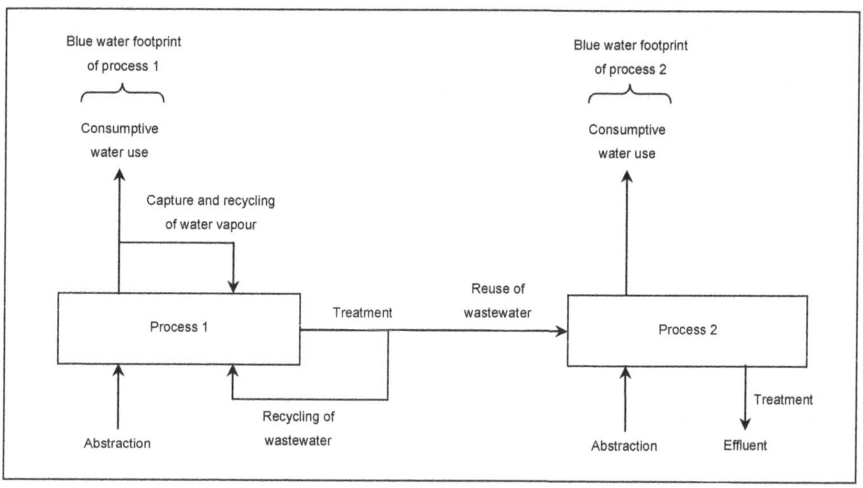

Figure 3.5 Blue water footprint accounting in the case of water recycling and reuse

Inter-basin water transfer

An inter-basin water transfer is the abstraction of water from a river basin A and move it – through pipelines, canals or bulk transport (for example, by lorry or ship) – to another river basin B. According to the blue water footprint definition, moving water away from a river basin is a blue water footprint within that basin, because it is 'consumptive water use'. The blue water footprint of the total transfer will be allocated to the beneficiaries of the water in the receiving river basin. Thus, processes in basin B that use water from another basin A have a blue water footprint located in basin A, the size of which is equal to the amount of water they receive plus the possible losses on the way. If the water users in the receiving river basin B return (part of) the used water to their own basin, we see that water is 'added' to the water resources in river basin B. This 'added' water may compensate for the blue water footprint of other users that have consumed water from basin B; in that sense one may argue that the inter-basin water transfer creates a 'negative blue water footprint' in the receiving river basin (insofar as the water does not evaporate and indeed adds to the water system of the receiving basin). The negative blue water footprint in basin B partly compensates the positive blue water footprint of other users in basin B. Note that it does not compensate for the blue water footprint in river basin A! When the goal is to assess the overall water footprint of humans in basin B, we recommend to include a possible 'negative blue water footprint' that exists as a result of real water transfer into the basin (provided that it indeed compensates for a positive blue water footprint in the basin in the same period). In the case of water footprint accounts for individual processes, products, consumers or producers, one should leave calculated negative blue water footprints out of the footprint accounts in order to make a clear separation between the discussion about the gross water footprint of a process, product, consumer or producer and the discussion about possible compensation. The issue of compensation (subtractability) is debatable and should be dealt with separately from the accounting phase. It has been argued that doing good in one basin (for example, through a negative blue water footprint in that basin) cannot compensate for the positive blue water footprint in another basin, since water depletion and resulting impacts in one place will not be solved by adding water somewhere else. In this case, adding a calculated negative blue water footprint to the calculated positive blue water footprint would result in a misleading figure. Read more on the impossibility to compensate a water footprint in one basin by adding water in another basin in Chapter 5 (Box 5.2).

3.3.2 Green water footprint

The green water footprint is an indicator of the human use of so-called green water. Green water refers to the precipitation on land that does not run off or recharge the groundwater but is stored in the soil or temporarily stays on top of the soil or vegetation. Eventually, this part of precipitation evaporates or transpires through plants. Green water can be made productive for crop growth (but not all green water can be taken up by crops, because there will always be evaporation from the soil and because not all periods of the year or areas are suitable for crop growth).

The green water footprint is the volume of rainwater consumed during the production process. This is particularly relevant for agricultural and forestry products (products based on crops or wood), where it refers to the total rainwater evapotranspiration (from fields and plantations) plus the water incorporated into the harvested crop or wood. The green water footprint in a process step is equal to:

$$WF_{proc,green} = GreenWaterEvaporation + GreenWaterIncorporation \quad [\text{volume/time}] \qquad (2)$$

The distinction between the blue and green water footprint is important because the hydrological, environmental and social impacts, as well as the economic opportunity costs of surface and groundwater use for production, differ distinctively from the impacts and costs of rainwater use (Falkenmark and Rockström, 2004; Hoekstra and Chapagain, 2008).

Green water consumption in agriculture can be measured or estimated with a set of empirical formulas or with a crop model suitable for estimating evapotranspiration based on input data on climate, soil and crop characteristics. In Section 3.3.4 we will present in more detail how one can estimate the green water footprint in crop growth.

3.3.3 Grey water footprint

The grey water footprint of a process step is an indicator of the degree of freshwater pollution that can be associated with the process step. It is defined as the volume of freshwater that is required to assimilate the load of pollutants based on natural background concentrations and existing ambient water quality standards. The grey water footprint concept has grown out of the recognition that the size of water pollution can be expressed in terms of the volume of water that is required to dilute pollutants such that they become harmless (Box 3.4).

The grey water footprint is calculated by dividing the pollutant load (*L*, in mass/time) by the difference between the ambient water quality standard for that pollutant (the maximum acceptable concentration c_{max}, in mass/volume) and its natural concentration in the receiving water body (c_{nat}, in mass/volume).

Box 3.4 *The history of the grey water footprint concept*

The grey water footprint refers to the volume of water that is required to assimilate waste, quantified as the volume of water needed to dilute pollutants to such an extent that the quality of the ambient water remains above agreed water quality standards. The idea of expressing water pollution in terms of a water volume needed to dilute the waste is not new. Falkenmark and Lindh (1974) proposed as a rule of thumb to reckon with a dilution factor of 10–50 times the wastewater flow. Postel et al (1996) applied a dilution factor for waste absorption of 28 litres per second per 1000 population. These generic dilution factors do not account for the sort of pollution and the level of treatment before disposal, but implicitly assume some average characteristics of human waste flows. Chapagain et al (2006b) proposed to make the dilution factor dependent on the type of pollutant and to use the ambient water quality standard for a certain pollutant as the criterion to quantify the dilution requirement. The term 'grey water footprint' was for the first time introduced by Hoekstra and Chapagain (2008) and defined as the pollutant load divided by the maximum acceptable concentration in the receiving water body. A bit later, it was recognized that the grey water footprint is better calculated as the pollutant load divided by the difference between the maximum acceptable and the natural background concentration (Hoekstra et al, 2009a). The work of the Water Footprint Network's grey water footprint working group (Zarate, 2010a) has further resulted in a number of refinements, including the recognition that the quality of intake water should be taken into account and the idea of a multi-tier approach in order to distinguish between different levels of detail in assessing a grey water footprint in the case of diffuse pollution.

Although the grey water footprint can be understood as a 'dilution water requirement', we prefer not to use that term since it appeared to cause confusion with some people who thought that the term implies we need to dilute pollutants instead of reduce their emission. This is, of course, not the meaning of the concept. The grey water footprint is an indicator of pollution and the less pollution the better. Treatment of wastewater before disposal will obviously result in a reduced grey water footprint, possibly down to zero.

Some recent studies that include quantification of grey water footprints include: Dabrowski et al (2009); Ercin et al (2009); Gerbens-Leenes and Hoekstra (2009); Van Oel et al (2009); Aldaya and Hoekstra (2010); Bulsink et al (2010); Chapagain and Hoekstra (2010); and Mekonnen and Hoekstra (2010a, b).

$$WF_{proc,grey} = \frac{L}{c_{max} - c_{nat}} \quad \text{[volume/time]} \qquad (3)$$

The natural concentration in a receiving water body is the concentration in the water body that would occur if there were no human disturbances in the catchment. For human-made substances that naturally do not occur in water, $c_{nat} = 0$. When natural concentrations are not known precisely but are estimated to be low, for simplicity one may assume $c_{nat} = 0$. This will, however, result in an underestimated grey water footprint when c_{nat} is actually not equal to zero.

One may ask why the natural concentration is used as a reference and not the actual concentration in the receiving water body. The reason is that the grey water footprint is an indicator of appropriated assimilation capacity. The assimilation capacity of a receiving water body depends on the difference between the maximum allowable and the natural concentration of a substance. If one would compare the maximum allowable concentration with the actual concentration of a substance, one would look at the *remaining* assimilation capacity, which is obviously changing all the time, as a function of the actual level of pollution at a certain time.

Grey water footprint calculations are carried out using ambient water quality standards for the receiving freshwater body, in other words, standards with respect to maximum allowable concentrations. The reason is that the grey water footprint aims to show the required ambient water volume to assimilate chemicals. Ambient water quality standards are a specific category of water quality standards. Other sorts of standards are, for example, drinking water quality standards, irrigation quality standards and emission (or effluent) standards. One should take care using ambient water quality standards. For one particular substance, the ambient water quality standard may vary from one to another water body. Besides, the natural concentration may vary from place to place. As a result, a certain pollutant load can result in one grey water footprint in one place and another grey water footprint in another place. This is reasonable, because the required water volume for assimilating a certain pollutant load will indeed be different depending on the difference between the maximum allowable and the natural concentration.

Although ambient water quality standards often exist in national or state legislation or have to be formulated by catchment and/or water body in the framework of national legislation or by regional agreement (like in the European Water Framework Directive – see EU, 2000), they do not exist for all substances and for all places. Most important is, of course, to specify which water quality standards and natural concentrations have been used in preparing a grey water footprint account.

Both ambient water quality standards and natural background concentrations vary for surface and groundwater bodies. Thresholds in groundwater are often based on requirements for drinking water, while maximum acceptable concentrations in surface waters are typically determined by ecological considerations. One could therefore propose to calculate the grey water footprint separately for surface and groundwater systems. The problem with doing so, however, is that groundwater generally ends up as surface water, so that for a pollutant load to groundwater one can better take the difference between water quality standard and natural background concentration for the most critical water body (either the groundwater system or the surface water system). For loads to the surface water system one can take the relevant data as for the surface water system. When it is precisely known which loads arrive (first) in the groundwater system and which loads in the surface water system, it makes sense to show two components of the grey water footprint: the grey groundwater footprint and the grey surface-water footprint.

A grey water footprint larger than zero does not automatically imply that ambient water quality standards are violated; it just shows that part of the assimilation capacity has been consumed already. As long as the calculated grey water footprint is smaller than the existing river flow or groundwater flow, there is still sufficient water to dilute the pollutants to a concentration below the standard. When the calculated grey water footprint is precisely equal to the ambient water flow, then the resultant concentration will be exactly at the standard. When the effluent contains a very high load of chemicals it may happen that the calculated grey water footprint exceeds the existing river flow or groundwater flow. In this case, pollution goes beyond the assimilation capacity of the receiving water body. The fact that the grey water footprint can be larger than the existing water flow illustrates that the grey water footprint does not show 'the polluted water volume' (because one would not be able to pollute a larger volume than the existing one). Instead, the grey water footprint is an indicator of the severity of water pollution, expressed in terms of the freshwater volume required to assimilate the existing load of pollutants.

The approach taken in grey water footprint accounting is the same as the so-called critical-load approach (Box 3.5). In both cases, the basic recognition is that the room for waste uptake of a water body is limited by the difference between the maximum and the natural concentration. The critical load refers to the situation where the room for waste uptake has been fully consumed. At the critical load, the grey water footprint will be equal to the available water flow, which is then required in full to dilute the chemicals down to acceptable concentrations.

Box 3.5 *The concept of critical load*

When the load into a flowing water body reaches a certain 'critical load', the grey water footprint will be equal to the run-off, which means that full run-off is appropriated for waste assimilation. The critical load (L_{crit}, in mass/time) is the load of pollutants that will fully consume the assimilation capacity of the receiving water body. It can be calculated by multiplying the run-off of the water body (*R*, in volume/time) by the difference between the maximum acceptable and natural concentration:

$$L_{crit} = R \times (c_{max} - c_{nat}) \quad \text{[mass/time]}$$

The concept of the 'critical load' is similar as the 'total maximum daily load' (TMDL) developed by the US Environmental Protection Agency (EPA, 2010a). The TMDL calculates the maximum amount of a pollutant allowed to enter a water body so that the water body will meet and continue to meet water quality standards for that particular pollutant and allocates that load to point and diffuse sources, which include both anthropogenic and natural background sources of the pollutant. Another concept closely related to the concept of 'critical load' is the concept of 'maximum permissible addition' (MPA), which is defined as the 'maximum permissible concentration' (MPC) minus the background concentration and thus equivalent to $c_{max} - c_{nat}$ (Crommentuijn et al, 2000).

Point sources of water pollution

In the case of point sources of water pollution, in other words, when chemicals are directly released into a surface water body in the form of a wastewater disposal, the load can be estimated by measuring the effluent volume and the concentration of a chemical in the effluent. More precisely: the pollutant load can be calculated as the effluent volume (*Effl*, in volume/time) multiplied by the concentration of the pollutant in the effluent (c_{effl}, in mass/volume) minus the water volume of the abstraction (*Abstr*, in volume/time) multiplied by the actual concentration of the intake water (c_{act}, in mass/volume). The grey water footprint can then be calculated as follows:

$$WF_{proc,grey} = \frac{L}{c_{max} - c_{nat}} = \frac{Effl \times c_{effl} - Abstr \times c_{act}}{c_{max} - c_{nat}} \quad \text{[volume/time]} \qquad (4)$$

The pollutant load *L* is thus defined as the load that comes on top of the load that was already contained in the receiving water body before interference by the activity considered. An example of the application of this equation in a concrete case is given in Appendix IV. Under most circumstances, the amount of chemicals discharged into a water body ($Effl \times c_{effl}$) will be equal or larger than the amount abstracted chemicals ($Abstr \times c_{act}$), so the load is positive. In exceptional circumstances (either because $c_{effl} < c_{act}$ or because $Effl < Abstr$), one could calculate a negative load, which has to be neglected for water footprint accounts (so one should account for a zero water footprint in such a case). The positive contribution made to the environment in the exceptional case of a 'negative load' is to be appreciated but should not be counted in the water footprint accounts, in order to separate the discussion of possible water footprint compensation from existing positive water footprints. Water footprint compensation or 'offsetting' is a debate in itself (see Box 5.2 in Chapter 5), which should be held and made explicit, not hidden in the accounts. It is further to be noted that, when water for a certain process is abstracted in catchment A while the effluent is discharged into catchment B, one should take $Abstr = 0$ for the calculation of the grey water footprint in catchment B.

When there is no consumptive water use, in other words, when the effluent volume equals the abstraction volume, above equation simplifies into:

$$WF_{proc,grey} = \frac{c_{effl} - c_{act}}{c_{max} - c_{nat}} \times Effl \qquad \text{[volume/time]} \qquad (5)$$

The factor that stands in front of *Effl* is the so-called 'dilution factor', which represents the number of times that the effluent volume has to be diluted with ambient water in order to arrive at the maximum acceptable concentration level. How this equation works out under a number of particular cases is discussed in Box 3.6.

Water recycling and reuse

From Equation 5, one can see that water recycling or water reuse will affect the grey water footprint. When – after treatment when necessary – water is fully recycled or reused for the same or another purpose, there is no effluent to the environment, so the grey water footprint will be zero. If after one time or a number of times of reuse, however, the water is still disposed into the environment, there will be a grey water footprint, related to the quality of the effluent of course.

Wastewater treatment

When wastewater is treated before it is disposed into the environment, this obviously lowers the concentration of pollutants in the final effluent, so it will lower the grey water footprint. It should be noted that the grey water footprint of a process depends on the quality of the effluent as it is finally disposed into the environment, not on the quality before treatment. Wastewater treatment

Box 3.6 *The grey water footprint in different cases of point-source pollution*

Let us consider the common case in which the effluent volume is equal (or close) to the abstraction volume.

- When $c_{effl} = c_{act}$ the associated grey water footprint is nil. This can easily be understood, because the concentration of the receiving water body will remain unchanged.
- When $c_{effl} = c_{max}$ the grey water footprint is equal to a certain fraction of the effluent volume. When, in addition, $c_{act} = c_{nat}$ then the grey water footprint is precisely equal to the effluent volume. One may ask: why there is a grey water footprint larger than zero when the effluent concentration meets the ambient water quality standard? The answer is that some of the capacity to assimilate pollutants has been consumed. Due to the effluent, the concentration of the chemical in the receiving water body has moved from c_{nat} in the direction of c_{max}. In the extreme case that all water in a river is withdrawn and returned as effluent with a concentration equal to c_{max}, then the full assimilation capacity of the river has been consumed, so the grey water footprint would be equal to the total river run-off.
- When $c_{effl} < c_{act}$ the calculated grey water footprint would be negative, which is explained by the fact that the effluent is cleaner than the intake water. 'Cleaning' when the river is actually still under natural conditions does not make much sense, because some background concentration is apparently natural. If, however, other activities have brought the natural concentration up already, cleaning actually contributes to bringing the ambient water quality back in the direction of natural conditions, so this positively contributes to water quality. However, calculated negative grey water footprints have to be ignored from the accounts, in order to separate the discussion on somebody's actual positive water footprint from the discussion on someone's possible role in terms of compensation. The issue of compensation or 'offsetting' of water footprints is discussed in Chapter 5 (Box 5.2).
- When $c_{max} = 0$ (the case of a complete ban of a highly persistent or toxic pollutant, for which also $c_{nat} = 0$), any effluent with a concentration larger than zero will create an infinitely large grey water footprint. This infiniteness corresponds to the absolute ban: absolutely unacceptable means the footprint goes sky high.
- The case of $c_{max} = c_{nat}$ creates an infinitely large grey water footprint as well, but this case will not occur, because setting standards equal to the natural concentration does not make sense and will normally not happen.

can bring the grey water footprint down to zero when the concentrations of pollutants in the effluent are equal to or lower than the concentrations in the water as it was abstracted. As a side remark, it is noted here that the process of wastewater treatment in itself will have a blue water footprint when evaporation takes place during the treatment process in open basins.

For thermal pollution, we can apply a similar approach as for pollution by chemicals. The grey water footprint is now calculated as the difference between the temperature of an effluent flow and the receiving water body (°C) divided by the maximum acceptable temperature increase (°C) times the effluent volume (volume/time):

$$WF_{proc,grey} = \frac{T_{effl} - T_{act}}{T_{\max} - T_{nat}} \times Effl \quad \text{[volume/time]} \tag{6}$$

The maximum acceptable temperature increase ($T_{max} - T_{nat}$) depends on the type of water and local conditions. If no local guideline is available, we recommend reckoning with a default value of 3°C (EU, 2006).

Diffuse sources of water pollution

Estimating the chemical load in the case of diffuse sources of water pollution is not as straightforward as in the case of point sources. When a chemical is applied on or put into the soil, as in the case of solid waste disposal or use of fertilizers or pesticides, it may happen that only a fraction seeps into the groundwater or runs off over the surface to a surface water stream. In this case, the pollutant load is the fraction of the total amount of chemicals applied (put on or into the soil) that reaches the groundwater or surface water. The amount of chemicals applied can be measured, but the fraction of applied chemicals that reaches the groundwater or surface water cannot be measured, since it enters the water in a diffuse way, so it is not clear where and when to measure. As a solution, one can measure the water quality at the outlet of a catchment, but different sources of pollution come together, so that the challenge becomes to apportion the measured concentrations to different sources. Therefore, it is common practice, and also recommended here, to estimate the fraction of applied chemicals that enter the water system by using simple or more advanced models. The simplest model is to assume that a certain fixed fraction of the applied chemicals finally reach the ground- or surface water:

$$WF_{proc,grey} = \frac{L}{c_{max} - c_{nat}} = \frac{\alpha \times Appl}{c_{max} - c_{nat}} \quad \text{[volume/time]} \tag{7}$$

The dimensionless factor α stands for the leaching-run-off fraction, defined as the fraction of applied chemicals reaching freshwater bodies. The variable *Appl*

Box 3.7 *Three-tier approach in estimating diffuse pollution loads*

A three-tier approach is recommended for estimating diffuse pollution loads, similar to the one of the Intergovernmental Panel on Climate Change (IPCC) for estimating greenhouse gas emissions (IPCC, 2006). From Tier 1 to 3, the accuracy increases but the feasibility decreases.

- Tier 1 uses a fixed fraction to translate data on the amount of chemicals applied to the soil to an estimate of the amount of chemicals that enter the groundwater or surface water system. The fraction is to be derived from existing literature and may depend on the chemical considered. This Tier 1 estimate will suffice as a first rough estimate but obviously excludes relevant factors such as soil type, agricultural practice, soil hydrology and interaction between different chemicals in the soil.
- Tier 2 applies standardized and simplified model approaches, which can be used based on widely available data (such as agricultural nutrient balances, soil loss data, basic hydrologic, petrologic and hydromorphologic information). These simple and standardized model approaches should be derived from widely accepted and validated models.
- Tier 3 uses sophisticated modelling techniques given that the available resources allow it and the chosen topic requires it. Whereas detailed mechanistic models of contaminant flows through soil are available, their complexity often renders them inappropriate for use in Tier 3 type modelling of the diffuse pollution load. However, validated empirical models driven by information on farm practices, and which use simplified soil and weather data are presently available for use in diffuse-load studies. Tier 3 studies should be used to further refine Tier 2 approaches.

represents the application of chemicals on or into the soil in a certain process (in mass/time). This model is the simplest, least detailed method to estimate the grey water footprint in the case of diffuse pollution; so it is recommended only as the default method, to be used if time does not allow a more detailed study. Higher levels of detail are possible; it is proposed to distinguish three levels of detail: from Tier 1 (the default method) to Tier 3 (the most detailed method). In Tiers 2 and 3, more specific data and advanced methods are to be used (Box 3.7).

The effect of evaporation on water quality

A specific form of 'pollution' can occur when water quality deteriorates as a result of evaporation. When a part of a water flow evaporates, the concentrations of chemicals in the remaining water flow will increase (because when water evaporates the chemicals in the water stay behind). Consider, as an example, the case of high salt concentrations in drainage water from irrigated fields. When there is continued irrigation with little drainage compared to the volume of water evaporating, the salts naturally contained in the irrigation water accumulate in the soil (since the water evaporates, not the salt). As a result, also the drainage water will have a relatively high salt content. One may call this 'pollution'. But obviously it is another sort of pollution than when humans add chemicals to the water, because in this case there is no addition of chemicals by humans, but the naturally present chemicals get concentrated by water evaporation. We can generalize this case to all cases where 'water is taken out of the system through evaporation'. It also happens, for instance, in artificial reservoirs where water evaporates and chemicals accumulate.

Increasing the chemical concentration in a water body by 'taking water out through evaporation while leaving the chemicals in' is effectively the same as adding a certain additional load. If one takes out X m^3 of pure water, the 'equivalent load' is X m^3 times the natural concentration in the water body (c_{nat} in mass per m^3). The 'equivalent load' of $X \times c_{nat}$ (expressed as a mass) is natural, but no longer embedded in natural water, because the water was taken away (it evaporated). This 'equivalent load' has to be assimilated by other natural water. The grey water footprint related to this 'equivalent load' can be calculated with the standard equation, whereby the grey water footprint is equal to the 'equivalent load' divided by the difference between the maximum and natural concentration (Equation 3). This grey water footprint will come on top of grey water footprints in the catchment related to real loads (in other words, chemical loads added by human activities).

Integration over time and different pollutants

Daily values for the grey water footprint can be added over the year to get annual values. When a waste flow concerns more than one form of pollution, as is generally the case, the grey water footprint is determined by the pollutant that is most critical, that is the one that is associated with the largest pollutant-specific grey water footprint. For the purpose of finding an overall indicator of water pollution, the grey water footprint based on the critical substance is sufficient. If one is interested in the pollutant-specific grey water footprints, one can of course report those values separately. For formulating response measures targeted at specific pollutants, this is of course very relevant. For the overall picture of pollution, however, showing the grey water footprint for the critical substance is good enough.

As a final note, it is observed here that grey water footprints are measured based on the (human-induced) loads that *enter into* freshwater bodies, not on the basis of the loads that can finally be measured in the river or groundwater flow at some *downstream* point. Since water quality evolves over time and in the course of the water flow as a result of natural processes, the load of a certain chemical at a downstream point can be distinctly different from the sum of the loads that once entered the stream (upstream). The choice to measure the grey water footprint at the point where pollutants enter the groundwater or surface water system has the advantage that it is relatively simple – because one does not need to model the processes that change water quality along the river – and safe – because water quality may improve along the flow of a river by decay processes – but it is unclear why one should take improved water quality downstream as an indicator instead of measuring the immediate impact of a load at the point where it enters the system. While the grey water footprint indicator thus does not account for natural processes that may improve water quality along the water flow, it does also not account for processes that consider the combined effect of pollutants, which may sometimes be greater than what one may expect on the basis of the concentrations of chemicals when considered separately. In the end, the grey water footprint strongly depends on ambient water quality standards (maximum acceptable concentrations), which is reasonable given the fact that such standards are set based on the best available knowledge about the possible harmful effects of chemicals including their possible interaction with other chemicals.

3.3.4 Calculation of the green, blue and grey water footprint of growing a crop or tree

Many products contain ingredients from agriculture or forestry. Crops are used for food, feed, fibre, fuel, oils, soaps, cosmetics and so on. Wood from trees and shrubs is used for timber, paper and fuel. Since the agricultural and forestry sectors are major water-consuming sectors, products that involve agriculture or forestry in their production system will often have a significant water footprint. For all those products it is relevant to particularly look into the water footprint of the process of growing the crop or tree. This section discusses the details of assessing the process water footprint of growing crops or trees. The method is applicable to both annual and perennial crops, where trees can be considered a perennial crop. In the following, the term 'crop' is used in a broad sense, thus also including 'trees' grown for the wood.

The total water footprint of the process of growing crops or trees (WF_{proc}) is the sum of the green, blue and grey components:

$$WF_{proc} = WF_{proc,green} + WF_{proc,blue} + WF_{proc,grey} \quad \text{[volume/mass]} \tag{8}$$

We will express all process water footprints in this section per unit of product, namely in water volume per mass. Usually we express process water footprints in agriculture or forestry as m^3/ton, which is equivalent to litre/kg.

The green component in the process water footprint of growing a crop or tree ($WF_{proc,green}$, m^3/ton) is calculated as the green component in crop water use (CWU_{green}, m^3/ha) divided by the crop yield (Y, ton/ha). The blue component ($WF_{proc,blue}$, m^3/ton) is calculated in a similar way:

$$WF_{proc,green} = \frac{CWU_{green}}{Y} \quad \text{[volume/mass]} \tag{9}$$

$$WF_{proc,blue} = \frac{CWU_{blue}}{Y} \quad \text{[volume/mass]} \tag{10}$$

Yields for annual crops can be taken as given in yield statistics. In the case of perennial crops, one should consider the average annual yield over the full lifespan of the crop. In this way, one accounts for the fact that the yield in the initial year of planting is low or zero, that yields are highest after some years and that yields often go down at the end of the life span of a perennial crop. Also for the crop water use, one needs to take the average annual crop water use over the life span of the crop.

The grey component in the water footprint of growing a crop or tree ($WF_{proc,grey}$, m^3/ton) is calculated as the chemical application rate to the field per hectare (AR, kg/ha) times the leaching-run-off fraction (α) divided by the maximum acceptable concentration (c_{max}, kg/m^3) minus the natural concentration for the pollutant considered (c_{nat}, kg/m^3) and then divided by the crop yield (Y, ton/ha).

$$WF_{proc,grey} = \frac{(\alpha \times AR)/(c_{\max} - c_{nat})}{Y} \quad \text{[volume/mass]} \tag{11}$$

The pollutants generally consist of fertilizers (nitrogen, phosphorus and so on), pesticides and insecticides. One has to consider only the 'waste flow' to freshwater bodies, which is generally a fraction of the total application of fertilizers or pesticides to the field. One needs to account for only the most critical pollutant, that is the pollutant where above calculation yields the highest water volume.

The green and blue components in crop water use (CWU, m^3/ha) are calculated by accumulation of daily evapotranspiration (ET, mm/day) over the complete growing period:

$$CWU_{green} = 10 \times \sum_{d=1}^{\text{lgp}} ET_{green} \quad \text{[volume/area]} \tag{12}$$

$$CWU_{blue} = 10 \times \sum_{d=1}^{\text{lgp}} ET_{blue} \quad \text{[volume/area]} \tag{13}$$

in which ET_{green} represents green water evapotranspiration and ET_{blue} blue water evapotranspiration. The factor 10 is meant to convert water depths in millimetres into water volumes per land surface in m^3/ha. The summation is done over the period from the day of planting (day 1) to the day of harvest (*lgp* stands for length of growing period in days). Since different crop varieties can have substantial differences in the length of the growing period, this factor can significantly influence the calculated crop water use. For permanent (perennial) crops and production forest, one should account for the evapotranspiration throughout the year. Besides, in order to account for differences in evapotranspiration over the full lifespan of a permanent crop or tree, one should look at the annual average of evapotranspiration over the full lifespan of the crop or tree. Suppose, for example, that a certain perennial crop has a lifespan of 20 years, while it gives a yield only from the sixth year on. In this case, the crop water use over the 20 years needs to be divided over the total yield over the 15 years of production. The 'green' crop water use represents the total rainwater evaporated from the field during the growing period; the 'blue' crop water use represents the total irrigation water evaporated from the field.

Evapotranspiration from a field can be either measured or estimated by means of a model based on empirical formulas. Measuring evapotranspiration is costly and unusual. Generally, one estimates evapotranspiration indirectly by means of a model that uses data on climate, soil properties and crop characteristics as input. There are many alternative ways to model ET and crop growth. One of the models frequently used is the EPIC model (Williams et al, 1989; Williams, 1995), also available in grid-based form (Liu et al, 2007). Another model is the CROPWAT model developed by the Food and Agriculture Organization of the United Nations (FAO, 2010b), which is based on the method described in Allen et al (1998). Yet another model is the AQUACROP model, specifically developed for estimating crop growth and ET under water-deficit conditions (FAO, 2010e).

The CROPWAT model offers two different options to calculate evapotranspiration: the 'crop water requirement option' (assuming optimal conditions) and the 'irrigation schedule option' (including the possibility to specify actual irrigation supply in time). We recommend to apply the second option whenever possible, because it is applicable for both optimal and non-optimal growing conditions and because it is more accurate (as the underlying model includes a dynamic soil water balance). A comprehensive manual for the practical use of the CROPWAT program is available online (FAO, 2010b). Appendix I summarizes how to use the 'crop water requirement option' to estimate green and blue water evapotranspiration under optimal conditions; it also summarizes the 'irrigation schedule option' that can be applied for all conditions. A practical example of the calculation of the process water footprint of growing a crop is given in Appendix II.

Estimating the green, blue and grey water footprints of growing a crop requires a large number of data sources (Box 3.8). In general it is always preferable to find local data pertaining to the crop field location. In many cases it is too laborious to collect location-specific data given the purpose of the assessment. If the purpose of the assessment allows a rough estimate, one can decide to work with data from nearby locations or with regional or national averages that may be more easily available.

In the above calculations, we have not yet accounted for the green and blue water incorporated into the harvested crop. One can find that component of the water footprint by simply looking at the water fraction of the harvested crop. For fruits this is typically in the range of 80–90 per cent of the wet mass, for vegetables often 90–95 per cent. The green-blue ratio in the water that is incorporated in the crop can be assumed equal to the ratio of CWU_{green} to CWU_{blue}. However, adding incorporated water to evaporated water will add little to the final water footprint number, because incorporated water is typically in the order of 0.1 per cent of the evaporated water, up to 1 per cent at most.

In this section we have looked into the calculation of the water footprint of growing a crop in the field. The blue water footprint calculated here refers to the evapotranspiration of irrigation water from the crop field only. It excludes the evaporation of water from artificial surface water reservoirs built for storing irrigation water and the evaporation of water from transport canals that bring the irrigation water from the place of abstraction to the field. Water storage and transport are two processes that precede the process of growing the crop in the field and have their own water footprint (Figure 3.6). The evaporation losses in these two preceding process steps can be very significant and should ideally be included when one is interested in the product water footprint of the harvested crop.

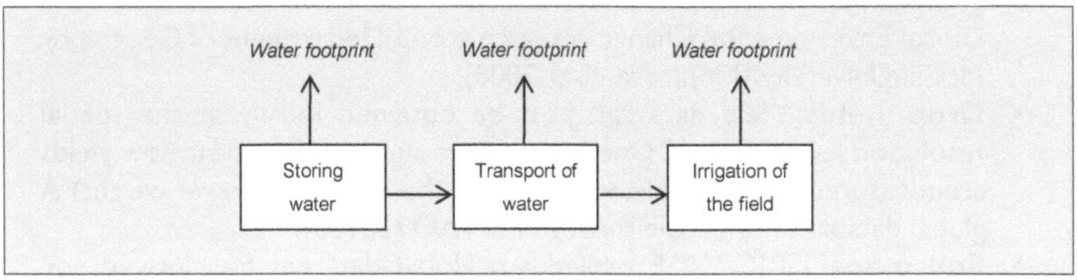

Figure 3.6 The subsequent processes in irrigation: storing water, transport of water, irrigation on the field. Each process step has its own water footprint

Box 3.8 *Data sources for the calculation of the water footprint of 'growing a crop'*

- **Climate data**: The calculation should be done using climate data from the nearest and most representative meteorological station(s) located near the crop field considered or within or near the crop-producing region considered. For regions with more than one climate station, one can make calculations for each station and weigh the outputs. The climate database CLIMWAT 2.0 (FAO, 2010a) provides the climatic data needed in the appropriate format required by the CROPWAT 8.0 model. The database does not provide data for specific years, but 30-year averages. Another source is LocClim 1.1 (FAO, 2005), which provides estimates of average climatic conditions at locations for which no observations are available. One can also use grid-based climate databases: monthly values of major climatic parameters with a spatial resolution of 30 arc minute can be obtained from CRU TS-2.1 through the CGIAR-CSI GeoPortal (Mitchell and Jones, 2005). The US National Climatic Data Centre provides daily climatic data for a large number of stations globally (NCDC, 2009). In addition, FAO provides through its GeoNetwork website long-term average precipitation and reference evapotranspiration with a spatial resolution of 10 arc minute (FAO, 2010g).
- **Crop parameters**: Crop coefficients and cropping pattern (planting and harvesting dates) can best be taken from local data. The crop variety and suitable growing period for a particular type of crop largely depends upon the climate and many other factors such as local customs, traditions, social structure, existing norms and policies. Therefore, the most reliable crop data are the data obtained from local agricultural research stations. Global databases that can be used are: Allen et al (1998, Tables 11–12), FAO (2010b), USDA (1994). FAO's online Global Information and Early Warning System (GIEWS) provides crop calendars for major crops for developing countries. One can access the zipped crop calendar images for each continent directly from the web (FAO, 2010f).
- **Crop maps**: Crop harvest areas and yields for 175 crops at 5 arc minute grid cell resolution are available from the website of the Land Use and Global Environmental Change research group, Department of Geography, McGill University (Monfreda et al, 2008).
- **Crop yields**: Yield data can best be obtained locally, at the spatial resolution level required. One has to make sure that it is clear how yields are measured (for example, what part of the crop, dry or wet weight). A global database is available through the FAO (2010d).
- **Soil maps**: ISRIC-WISE provides a global data set for derived soil properties both at 5 arc minute and 30 arc minute resolution (Batjes, 2006). In addition, the FAO GeoNetwork website provides maximum available soil moisture data at 5 arc minute resolution (FAO, 2010h). When applying the 'irrigation schedule option' in the CROPWAT model, one needs soil data; if no soil data are available we advise to choose 'medium soil' as a default.

- **Irrigation maps**: The Global Map of Irrigation Areas (GMIA) version 4.0.1 (Siebert et al, 2007) with a spatial resolution of 5 arc minute defines areas equipped for irrigation. Irrigation maps for 26 major crops both at 5 and 30 arc minute resolutions can be obtained from University of Frankfurt website (Portmann et al, 2008, 2010). These data also provide rain-fed crop growing areas for the same 26 crops.
- **Fertilizer application rates**: Preferably one uses local data. A useful global database is FertiStat (FAO, 2010c). The International Fertilizer Association (IFA, 2009) provides annual fertilizer consumption per country. Heffer (2009) provides fertilizer use per crop for major crop types and major countries.
- **Pesticides application rates**: Preferably one uses local data. The National Agricultural Statistics Service (NASS, 2009) provides an online database for the US with chemical use per crop. CropLife Foundation (2006) provides a database on pesticides use in the USA. Eurostat (2007) gives data for Europe.
- **Leaching-run-off fraction**: No databases available. One will have to work with experimental data from field studies and make rough assumptions. One can assume 10 per cent for nitrogen fertilizers, following Chapagain et al (2006b).
- **Ambient water quality standards**: Preferably use local standards as regulated in legislation. This information is available for some parts of the world, such as for the European Union (EU, 2008), the US (EPA, 2010b), Canada (Canadian Council of Ministers of the Environment, 2010), Australia/New Zealand (ANZECC and ARMCANZ, 2000), China (Chinese Ministry of Environmental Protection, 2002), Japan (Japanese Ministry of the Environment, 2010), Austria (Austrian Federal Ministry of Agriculture, Forestry, Environment and Water Management, 2010), Brazil (CONAMA, 2005), South Africa (South African Department of Water Affairs and Forestry, 1996), Germany (LAWA-AO, 2007) and the UK (UKTAG, 2008). A compilation can be found in MacDonald et al (2000). If no ambient water quality standards are available and the water body is to be suitable for drinking, one can decide to apply drinking water standards. See, for example, EU (2000) and EPA (2005).
- **Natural concentrations in receiving water bodies**: In more or less pristine rivers, one can assume that natural concentrations are equal to the actual concentrations and thus rely on long-term daily or monthly averages as measured in a nearby measuring station. For disturbed rivers, one will have to rely on historical records or model studies. For some parts of the world good studies are available; for the US see, for example, Clark et al (2000) and Smith et al (2003); for Austria see Austrian Federal Ministry of Agriculture, Forestry, Environment and Water Management (2010); for Germany see LAWA-AO (2007). As a reference, a global database on actual (not natural) concentrations is available through UNEP (2009). When no information is available, assume the natural concentration according to the best estimate or to be zero.
- **Actual concentration of the intake water:** A global database on actual concentrations is available through UNEP (2009).

3.4 Water footprint of a product

3.4.1 Definition

The water footprint of a product is defined as the total volume of fresh water that is used directly or indirectly to produce the product. It is estimated by considering water consumption and pollution in all steps of the production chain.[1] The accounting procedure is similar to all sorts of products, be it products derived from the agricultural, industrial or service sector. The water footprint of a product breaks down into a green, blue and grey component. An alternative term for the water footprint of a product is its 'virtual-water content', but the meaning of the latter term is narrower (Box 3.9).

Box 3.9 *Terminology: Water footprint, virtual-water content, embedded water*

The water footprint of a product is similar to what in other publications has been called alternatively the 'virtual-water content' of the product or the product's embedded, embodied, exogenous or shadow water (Hoekstra and Chapagain, 2008). The terms virtual-water content and embedded water, however, refer to the water volume embodied in the product alone, while the term 'water footprint' refers not only to the volume, but also to the sort of water that was used (green, blue, grey) and to when and where the water was used. The water footprint of a product is thus a multidimensional indicator, whereas 'virtual-water content' or 'embedded water' refer to a volume alone. We recommend using the term 'water footprint' because of its broader scope. The volume is just one aspect of water use; place and timing of water use and type of water used are as important. Besides, the term 'water footprint' can also be used in a context where we speak about the water footprint of a consumer or producer. It would sound strange to speak about the virtual-water content of a consumer or producer. We use the term 'virtual water' in the context of international (or interregional) virtual-water flows. If a nation (region) exports/imports a product, it exports/imports water in virtual form. In this context one can speak about virtual-water export or import, or more general about virtual-water flows or trade.

In the case of agricultural products, the water footprint is generally expressed in terms of m^3/ton or litres/kg. In many cases, when agricultural products are countable, the water footprint can also be expressed as a water volume per piece. In the case of industrial products, the water footprint can be expressed in terms of m^3/US$ or water volume per piece. Other ways to express a product water footprint are for example water volume/kcal (for food products in the context of diets) or water volume/joule (for electricity or fuels).

3.4.2 Schematization of the production system into process steps

In order to estimate the water footprint of a product, one will have to start by understanding the way a product is produced. For that reason, one will have to identify the 'production system'. A production system consists of sequential 'process steps'. A (simplified) example of the production system of a cotton shirt is: cotton growth, harvesting, ginning, carding, knitting, bleaching, dying, printing, finishing. Given the fact that many products require multiple inputs, it often happens that multiple process steps precede one next process step. In such a case we will not have a linear chain of process steps, but rather a 'product tree'. A (simplified) example of a product tree is: produce feed and all sorts of other inputs necessary in intensive livestock farming, raise the animals and finally produce meat. Since production systems often produce more than one final product – for example, cows can deliver milk as well as meat and leather – even the metaphor of a product tree is insufficient. In reality production systems are complex networks of linked processes, in many cases even circular.

For estimating the water footprint of a product, one will have to schematize the production system into a limited number of linked process steps. Besides, when one intends to go beyond a very superficial analysis based on global averages, one will have to specify the steps in time and space, which means that one will have to trace the origin of the (inputs of the) product. In the cotton shirt example above, cotton growth may happen in one place (China), while manufacturing may happen in another place (Malaysia) and consumption in yet another place (Germany). Production circumstances and process characteristics will differ from place to place, so that place of production will influence the size and colour of the water footprint. Besides, in the end one may like to be able to geographically map the water footprint of a final product, so that is another reason to keep track of place.

Schematization of a production system into distinct process steps inevitably requires assumptions and simplifications. Particularly relevant is the truncation problem mentioned in Chapter 2. Theoretically, because many production systems contain circular components, one could infinitely keep on tracing back inputs through the network of linked process steps. In practice, one will have to stop the analysis at those points where additional work will not add more significant information for the purpose of the analysis.

Production system diagrams for agricultural products can be found for example in FAO (2003) and Chapagain and Hoekstra (2004). For industrial products, one can generally relatively easily construct a production system diagram based on publicly available data sources. Better, of course, is to seek information on which process steps are taken in the actual supply chain of the product considered. This requires the tracing of all product ingredients.

3.4.3 Calculation of a product water footprint

The water footprint of a product can be calculated in two alternative ways: with the chain-summation approach or the stepwise accumulative approach. The former can be applied for particular cases only; the latter is the generic approach.

The chain-summation approach

This approach is simpler than the one that will be discussed next, but can be applied only in the case where a production system produces one output product (Figure 3.7). In this particular case, the water footprints that can be associated with the various process steps in the production system can all be fully attributed to the product that results from the system.

In this simple production system, the water footprint of product p (volume/mass) is equal to the sum of the relevant process water footprints divided by the production quantity of product p:

$$WF_{prod}[p] = \frac{\sum_{s=1}^{k} WF_{proc}[s]}{P[p]} \quad \text{[volume/mass]} \qquad (14)$$

in which $WF_{proc}[s]$ is the process water footprint of process step s (volume/time), and $P[p]$ the production quantity of product p (mass/time). In practice, simple production systems with only one output product rarely exist, thus a more generic way of accounting is necessary, one that can distribute the water used

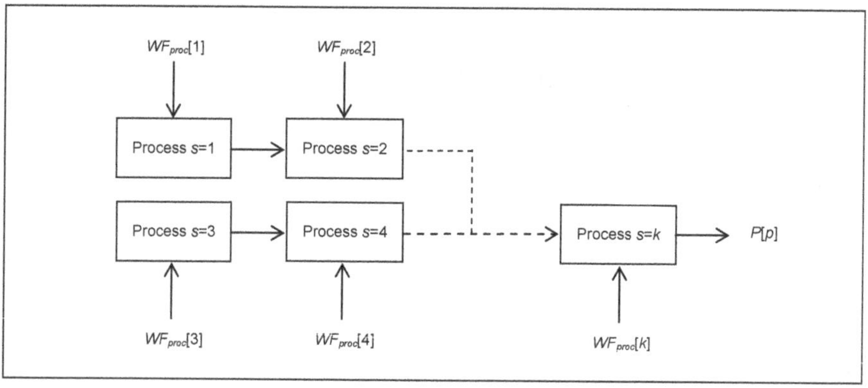

Figure 3.7 Schematization of the production system to produce product *p* into *k* process steps. Some steps are in series, others are parallel. The water footprint of output product *p* is calculated as the sum of the process water footprints of the processes that constitute the production system. Note: this simplified scheme presupposes that *p* is the only output product following from the production system

throughout a production system to the various output products that follow from that system without double counting.

The stepwise accumulative approach

This approach is a generic way of calculating the water footprint of a product based on the water footprints of the input products that were necessary in the last processing step to produce that product and the process water footprint of that processing step. Suppose we have a number of input products when making one output product. In this case we can get the water footprint of the output product by simply summing the water footprints of the input products and add the process water footprint. Suppose another case where we have one input product and a number of output products. In this case, one needs to distribute the water footprint of the input product to its separate products. This can be done proportionally to the value of the output products. It could also be done proportionally to the weight of the products, but this would be less meaningful. Finally, consider the most generic case (Figure 3.8). We want to calculate the water footprint of a product p, which is being processed from y input products. The input products are numbered from $i=1$ to y. Suppose that processing of the y input products results in z output products. We number the output products from $p=1$ to z.

If there is some water use involved during processing, the process water footprint is added to the water footprints of the input products before the total

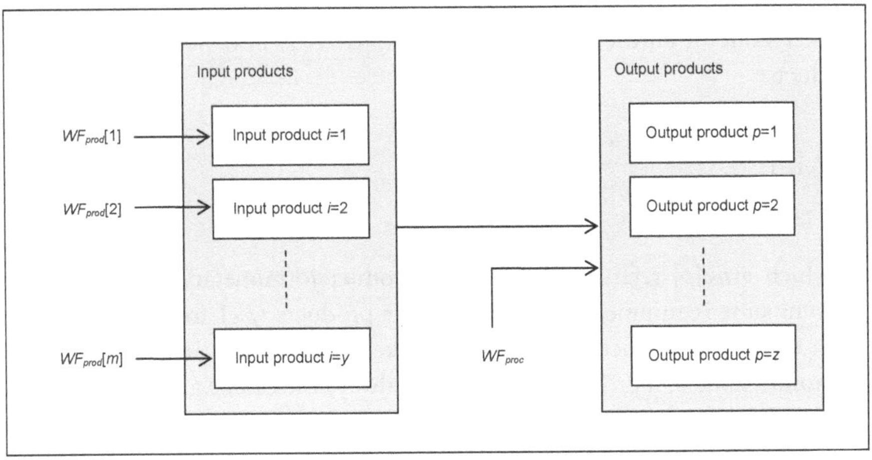

Figure 3.8 Schematization of the last process step in the production system to produce product *p*. The water footprint of output product *p* is calculated based on the water footprints of the input products and the process water footprint when processing the inputs into the outputs

is distributed over the various output products. The water footprint of output product p is calculated as:

$$WF_{prod}[p] = \left(WF_{proc}[p] + \sum_{i=1}^{y} \frac{WF_{prod}[i]}{f_p[p,i]} \right) \times f_v[p] \quad \text{[volume/mass]} \tag{15}$$

in which $WF_{prod}[p]$ is the water footprint (volume/mass) of output product p, $WF_{prod}[i]$ the water footprint of input product i and $WF_{proc}[p]$ the process water footprint of the processing step that transforms the y input products into the z output products, expressed in water use per unit of processed product p (volume/mass). Parameter $f_p[p,i]$ is a so-called 'product fraction' and parameter $f_v[p]$ is a 'value fraction'. Both will be defined below. Be aware that in the equation the process water footprint is to be taken in terms of water volume per unit of *processed* product; when the process water footprint is given per unit of a specific *input* product, the given volume needs to be divided by the product fraction for that input product.

The product fraction of an output product p that is processed from an input product i ($f_p[p,i]$, mass/mass) is defined as the quantity of the output product ($w[p]$, mass) obtained per quantity of input product ($w[i]$, mass):

$$f_p[p,i] = \frac{w[p]}{w[i]} \quad [-] \tag{16}$$

The value fraction of an output product p ($f_v[p]$, monetary unit/monetary unit) is defined as the ratio of the market value of this product to the aggregated market value of all the outputs products (p=1 to z) obtained from the input products:

$$f_v[p] = \frac{price[p] \times w[p]}{\sum_{p=1}^{z} (price[p] \times w[p])} \quad [-] \tag{17}$$

in which *price*[p] refers to the price of product p (monetary unit/mass). The denominator is summed over the z output products (p=1 to z) that originate from the input products. Note that we take 'price' here as an indicator of the economic value of a product, which is not always the case, e.g. when there is no market for a product or when the market is distorted. Of course one can best take the real economic value.

Note that in a simple case, where we process just one input product into one output product, the calculation of the water footprint of the output product becomes rather simple:

$$WF_{prod}[p] = WF_{proc}[p] + \frac{WF_{prod}[i]}{f_p[p,i]} \quad \text{[volume/mass]} \tag{18}$$

In order to calculate the water footprint of the final product in a production system, one can best start calculating the water footprints of the most original resources (where the supply chain starts) and then calculate, step-by-step, the water footprints of the intermediate products, until one can calculate the water footprint of the final product. The first step is always to obtain the water footprints of the input products and the water used to process them into the output product. The total of these components is then distributed over the various output products, based on their product fraction and value fraction.

A practical example of the calculation of the water footprint of a crop product is given in Appendix III.

Product fractions can best be taken from the literature available for a specific production process. Product fractions are often in a rather narrow range, but sometimes the amount of output product per unit of input product really depends on the precise process applied. In that case it is important to know which type of process is being applied in the case considered. For crop and livestock products, product fractions can be found in FAO (2003) and in Chapagain and Hoekstra (2004, vol 2). Value fractions fluctuate over the years depending on price developments. In order to avoid a large effect of the price fluctuation on the outcome of the water footprint calculations, we recommend estimating value fractions based on the average price over at least a five-year period. Value fractions for a large range of crop and livestock products are reported in Chapagain and Hoekstra (2004). We recommend, however, to look first for data that link to the actual case considered before taking default values from literature. The process water footprint in a certain process step may vary depending on the type of method applied (for example, wet or dry milling, dry or wet cleaning, closed cooling system or open cooling system with water evaporating). For many processes, one can find some estimates on water withdrawals in literature, but not on consumptive water use. General data on pollution per process are also scarce; they will strongly vary from place to place as well, so using general estimates would be very crude. One will have to look for the data at the source, namely the producers and factories.

3.5 Water footprint of a consumer or group of consumers

3.5.1 Definition

The water footprint of a consumer is defined as the total volume of freshwater consumed and polluted for the production of the goods and services used by the consumer. The water footprint of a group of consumers is equal to the sum of the water footprints of the individual consumers.

3.5.2 Calculation

The water footprint of a consumer (WF_{cons}) is calculated by adding the direct water footprint of the individual and his/her indirect water footprint:

$$WF_{cons} = WF_{cons,dir} + WF_{cons,indir} \quad \text{[volume/time]} \qquad (19)$$

The direct water footprint refers to the water consumption and pollution that is related to water use at home or in the garden. The indirect water footprint refers to the water consumption and pollution of water that can be associated with the production of the goods and services used by the consumer. It refers to the water that was used to produce for example the food, clothes, paper, energy and industrial goods consumed. The indirect water use is calculated by multiplying all products consumed by their respective product water footprint:

$$WF_{cons,indir} = \sum_p \left(C[p] \times WF^*_{prod}[p] \right) \quad \text{[volume/time]} \qquad (20)$$

$C[p]$ is consumption of product p (product units/time) and $WF^*_{prod}[p]$ the water footprint of this product (water volume/product unit). The set of products considered refers to the full range of final consumer goods and services. The water footprint of a product is defined and calculated as described in the previous section.

The total volume of p consumed will generally originate from different places x. The average water footprint of a product p consumed is calculated as:

$$WF^*_{prod}[p] = \frac{\sum_x \left(C[x,p] \times WF_{prod}[x,p] \right)}{\sum_x C[x,p]} \quad \text{[volume/product unit]} \qquad (21)$$

where $C[x,p]$ is consumption of product p from origin x (product units/time) and $WF_{prod}[x,p]$ the water footprint of product p from origin x (water volume/

product unit). Depending on the preferred level of detail of analysis, one can trace the origin of the products consumed with more or less precision. If one cannot or does not want to trace the origins of the products consumed, one will have to rely on either global or national average estimates of the water footprints of the products consumed. If, however, one is prepared to trace the origin of products, one can estimate the product water footprints with a high level of spatial detail (see the alternative levels of spatiotemporal explication in water footprint accounting as described in Chapter 2). Preferably the consumer knows per product how much he or she consumes from various origins. If the consumer does not know that, one can assume that the variation in origin equals the variation in origin as available on the national market for that product. The value of $WF^{*}_{prod}[p]$ can then be calculated with the formula as will be introduced in Section 3.7.3.

The water footprints of final private goods and services are exclusively allocated to the consumer of the private good. The water footprints of public or shared goods and services are allocated to consumers based on the share that each individual consumer takes.

3.6 Water footprint within a geographically delineated area

3.6.1 Definition

The water footprint within a geographic area is defined as the total freshwater consumption and pollution within the boundaries of the area. It is crucial to clearly define the boundaries of the area considered. The area can be a catchment area, a river basin, a province, state or nation or any other hydrological or administrative spatial unit.

3.6.2 Calculation

The water footprint within a geographically delineated area (WF_{area}) is calculated as the sum of the process water footprints of all water using processes in the area:

$$WF_{area} = \sum_{q} WF_{proc}[q] \quad \text{[volume/time]} \qquad (22)$$

where $WF_{proc}[q]$ refers to the water footprint of a process q within the geographically delineated area. The equation sums over all water-consuming or polluting processes taking place in the area.

The export of real water out of an area, as in the case of an inter-basin transfer, will be counted as a process water footprint in the area from which the water is exported.

From the perspective of water resources protection within a certain area – particularly when the area is water-scarce – it is interesting to know how much water is used in the area to produce export products and how much water is imported in virtual form (in the form of water-intensive products) so that they do not need to be produced within the area. In other words, it is interesting to know the 'virtual-water balance' of an area. The virtual-water balance of a geographically delineated area over a certain time period is defined as the net import of virtual water over this period ($V_{i,net}$), which is equal to the gross import of virtual water (V_i) minus the gross export (V_e):

$$V_{i,net} = V_i - V_e \quad \text{[volume/time]} \tag{23}$$

A positive virtual-water balance implies net inflow of virtual water to the area from other areas. A negative balance means net outflow of virtual water. The gross virtual-water import is interesting in the sense that importing virtual water saves water within the area considered. The gross virtual-water export is interesting in the sense that it refers to a water footprint in the area related to consumption by people living outside the area. Virtual-water imports and exports can be calculated following the same approach as specifically discussed for the case of nations in Section 3.7.3.

3.7 National water footprint accounting

3.7.1 The national water footprint accounting scheme

Full national water footprint accounts are obtained by combining the accounts on the 'water footprint of national consumption' (consumer accounts as introduced in Section 3.5) and the accounts on the 'water footprint within a nation' (area accounts as introduced in Section 3.6) in one comprehensive scheme. Figure 3.9 shows a visual representation of the national water footprint accounting scheme as was introduced by Hoekstra and Chapagain (2008).

Traditional national water use accounts only refer to the water withdrawal within a country. They do not distinguish between water use for making products for domestic consumption and water use for producing export products. They also exclude data on water use outside the country to support national consumption. In addition, they include blue water use only, excluding green and grey water. In order to support a broader sort of analysis and better inform decision-making, the traditional national water use accounts need to be extended.

The water footprint of the consumers in a nation ($WF_{cons,nat}$) has two components: the internal water footprint and the external water footprint.

$$WF_{cons,nat} = WF_{cons,nat,int} + WF_{cons,nat,ext} \quad \text{[volume/time]} \tag{24}$$

The internal water footprint of national consumption ($WF_{cons,nat,int}$) is defined as the use of domestic water resources to produce goods and services consumed by the national population. It is the sum of the water footprint within the nation ($WF_{area,nat}$) minus the volume of virtual-water export to other nations insofar as related to the export of products produced with domestic water resources ($V_{e,d}$):

$$WF_{cons,nat,int} = WF_{area,nat} - V_{e,d} \quad \text{[volume/time]} \tag{25}$$

The external water footprint of national consumption ($WF_{cons,nat,ext}$) is defined as the volume of water resources used in other nations to produce goods and services consumed by the population in the nation considered. It is equal to the virtual-water import into the nation (V_i) minus the volume of virtual-water export to other nations as a result of re-export of imported products ($V_{e,r}$):

$$WF_{cons,nat,ext} = V_i - V_{e,r} \quad \text{[volume/time]} \tag{26}$$

The virtual-water export (V_e) from a nation consists of exported water of domestic origin ($V_{e,d}$) and re-exported water of foreign origin ($V_{e,r}$):

$$V_e = V_{e,d} + V_{e,r} \quad \text{[volume/time]} \tag{27}$$

The virtual-water import into a nation will partly be consumed, thus constituting the external water footprint of national consumption ($WF_{cons,nat,ext}$), and partly be re-exported ($V_{e,r}$):

$$V_i = WF_{cons,nat,ext} + V_{e,r} \quad \text{[volume/time]} \tag{28}$$

The sum of V_i and $WF_{area,nat}$ is equal to the sum of V_e and $WF_{cons,nat}$. This sum is called the virtual-water budget (V_b) of a nation.

$$V_b = V_i + WF_{area,nat} = V_e + WF_{cons,nat} \quad \text{[volume/time]} \tag{29}$$

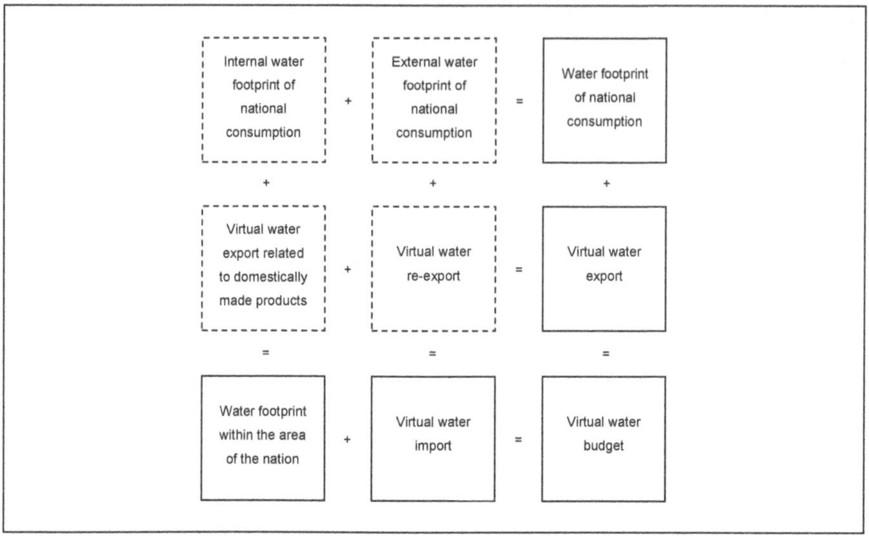

Figure 3.9 The national water footprint accounting scheme. The accounting scheme shows the various balances that hold for the water footprint related to national consumption ($WF_{cons,nat}$), the water footprint within the area of the nation ($WF_{area,nat}$), the total virtual-water export (V_e) and the total virtual-water import (V_i).

3.7.2 Calculation of the water footprint within a nation

The water footprint within a nation ($WF_{area,nat}$, volume/time) is defined as the total freshwater volume consumed or polluted within the territory of the nation. It can be calculated following the method described in Section 3.6:

$$WF_{area,nat} = \sum_{q} WF_{proc}[q] \quad \text{[volume/time]} \tag{30}$$

where $WF_{proc}[q]$ refers to the water footprint of process q within the nation that consumes or pollutes water. The equation sums over all water consuming or polluting processes taking place in the nation. Process water footprints are expressed here in volume/time.

3.7.3 Calculation of the water footprint of national consumption

The water footprint of national consumption ($WF_{cons,nat}$) can be calculated through two alternative approaches: the top-down and the bottom-up approach.

Top-down approach

In the top-down approach, the water footprint of national consumption ($WF_{cons,nat}$, volume/time) is calculated as the water footprint within the nation ($WF_{area,nat}$) plus the virtual-water import (V_i) minus the virtual-water export (V_e):

$$WF_{cons,nat} = WF_{area,nat} + V_i - V_e \quad \text{[volume/time]} \qquad (31)$$

The gross virtual-water import is calculated as:

$$V_i = \sum_{n_e}\sum_{p}\left(T_i[n_e, p] \times WF_{prod}[n_e, p]\right) \quad \text{[volume/time]} \qquad (32)$$

in which $T_i[n_e,p]$ represents the imported quantity of product p from exporting nation n_e (product units/time) and $WF_{prod}[n_e,p]$ the water footprint of product p as in the exporting nation n_e (volume/product unit). If further details are not available, one can assume that a product is produced in the exporting country. One can thus take the average product water footprint as in the exporting country. If one knows the location of origin within the exporting country, one can take the location-specific product water footprint. When a product is imported from a country that does not produce the product and when further information about the real origin is lacking, one can assume the global average product water footprint for that import flow. Ideally, for each product import, one takes the product water footprint as measured along the actual supply chain of the product, but in practice this is possibly doable on a case-by-case basis (as shown by Chapagain and Orr (2008) in a study of UK's water footprint), but not in generic sense for all imports into a country. Obviously, one needs to specify the specific assumptions taken in this respect.

The gross virtual-water export is calculated as:

$$V_e = \sum_{p} T_e[p] \times WF^*_{prod}[p] \quad \text{[volume/time]} \qquad (33)$$

in which $T_e[p]$ represents the quantity of product p exported from the nation (product units/time) and $WF^*_{prod}[p]$ the average water footprint of the exported product p (volume/product unit). The latter is estimated as:

$$WF^*_{prod}[p] = \frac{P[p] \times WF_{prod}[p] + \sum_{n_e}\left(T_i[n_e, p] \times WF_{prod}[n_e, p]\right)}{P[p] + \sum_{n_e} T_i[n_e, p]} \quad \text{[volume/product unit]} \qquad (34)$$

in which $P[p]$ represents the production quantity of product p in the nation, $T_i[n_e,p]$ the imported quantity of product p from exporting nation n_e, $WF_{prod}[p]$ the water footprint of product p when produced in the nation considered and

$WF_{prod}[n_e,p]$ the water footprint of product p as in the exporting nation n_e. The assumption made here is that export originates from domestic production and imports according to their relative volumes.

Bottom-up approach

The bottom-up approach is based on the method of calculating the water footprint of a group of consumers (Section 3.5). The group of consumers consists of the inhabitants of a nation. The water footprint of national consumption is calculated by adding the direct and indirect water footprints of consumers within the nation:

$$WF_{cons,nat} = WF_{cons,nat,dir} + WF_{cons,nat,indir} \quad \text{[volume/time]} \qquad (35)$$

The direct water footprint refers to consumption and pollution of water due to water use by consumers at home or in the garden. The indirect water footprint of consumers refers to the water use by others to make the goods and services consumed. It refers to the water that was used to produce for example the food, clothes, paper, energy and industrial goods consumed. The indirect water footprint is calculated by multiplying all products consumed by the inhabitants of the nation by their respective product water footprint:

$$WF_{cons,nat,indir} = \sum_{p} \left(C[p] \times WF^{*}_{prod}[p] \right) \quad \text{[volume/time]} \qquad (36)$$

$C[p]$ is consumption of product p by consumers within the nation (product units/time) and $WF^{*}_{prod}[p]$ the water footprint of this product (volume/product unit). The set of products considered refers to the full range of final consumer goods and services. The volume of p consumed in a nation will generally partly originate from the nation itself and partly from other nations. The average water footprint of a product p consumed in a nation is estimated by applying the same assumption that was used in the top-down approach:

$$WF^{*}_{prod}[p] = \frac{P[p] \times WF_{prod}[p] + \sum_{n_e} \left(T_i[n_e,p] \times WF_{prod}[n_e,p] \right)}{P[p] + \sum_{n_e} T_i[n_e,p]} \quad \text{[volume/product unit]} \qquad (37)$$

The assumption is that consumption originates from domestic production and imports according to their relative volumes.

The bottom-up versus the top-down approach

The bottom-up and top-down calculations theoretically result in the same figure, provided that there is no product stock change over a year. The top-down calculation can theoretically give a slightly higher (lower) figure if the stocks of water-intensive products increase (decrease) over the year. The reason is that the top-down approach presupposes a balance: $WF_{area,nat}$ plus V_i becomes $WF_{cons,nat}$ plus V_e. This is an approximation only, because, to be more precise: $WF_{area,nat}$ plus V_i becomes $WF_{cons,nat}$ plus V_e plus virtual-water stock increase. Another drawback of the top-down approach is that there can be delays between the moment of water use for production and the moment of trade. For instance, in the case of trade in livestock products, this may happen: beef or leather products traded in one year originate from livestock raised and fed in previous years. Part of the water virtually embedded in beef or leather refers to water that was used to grow feed crops in previous years. As a result of this, the balance presumed in the top-down approach will hold over a period of a few years, but not necessarily over a single year.

Next to theoretical differences between the two approaches, differences can result from the use of different types of data as inputs of the calculations. The bottom-up approach depends on the quality of consumption data, while the top-down-approach relies on the quality of trade data. When the different databases are not consistent with one another, the results of both approaches will differ. In one particular type of case, the outcome of the top-down can be very vulnerable to relatively small errors in the input data. This happens when the import and export of a country are large relative to its domestic production, which is typical for relatively small nations specialized in trade. This has been shown in a case study for the Netherlands (Van Oel et al, 2009). In this case, the water footprint of national consumption calculated with the top-down approach, will be sensitive to the import and export data used. Relative small errors in the estimates of virtual-water import and export translate into a relatively large error in the water footprint estimate. In such a case, the bottom-up approach will yield a more reliable estimate than the top-down approach. In nations where trade is relatively small compared to domestic production, the reliability of the outcomes of both approaches will depend on the relative quality of the databases used for each approach.

External water footprint of national consumption

With either the top-down or bottom-up approach one can calculate the total water footprint of national consumption ($WF_{cons,nat}$). With the top-down approach, one can calculate the virtual-water import into a country (V_i). Earlier, in Section 3.7.2, we have seen how one can calculate the water footprint within a nation ($WF_{area,nat}$). Based on these data, the external water footprint of national consumption ($WF_{cons,nat,ext}$) can be calculated as:

$$WF_{cons,nat,ext} = \frac{WF_{cons,nat}}{WF_{area,nat} + V_i} \times V_i \quad \text{[volume/time]} \tag{38}$$

This formula can be applied separately for the category of agricultural products (crop and livestock products) and for the category of the industrial products. The formula says that only a fraction of the gross virtual-water import can be said to be the external water footprint of national consumption and that this fraction is equal to the portion of the virtual water budget (sum of water footprint within the nation and virtual-water import) that is to be attributed to national consumption.[2] The other portion of the virtual water budget is exported and is therefore not part of the water footprint of national consumption.

The external water footprint of national consumption can be estimated by export nation n_e and product p by assuming that the national ratio between the external water footprint and the total virtual-water import applies to all partner nations and imported products:[3]

$$WF_{cons,nat,ext}[n_e, p] = \frac{WF_{cons,nat,ext}}{V_i} \times V_i[n_e, p] \quad \text{[volume/time]} \tag{39}$$

It happens that products are imported from nations in which they are not produced. For those products one will have to trace the origin country further back. For some product groups, world production is concentrated in specific regions. For these products one can roughly estimate the ultimate place of origin based on world production data. This means that one distributes the water footprint in a non-producing nation over producing nations according to the distribution of the world production.

3.7.4 Water savings related to trade

The national water saving S_n (volume/time) of a nation as a result of trade in product p is defined as:

$$S_n[p] = (T_i[p] - T_e[p]) \times WF_{prod}[p] \quad \text{[volume/time]} \tag{40}$$

where $WF_{prod}[p]$ is the water footprint (volume/product unit) of product p in the nation considered, $T_i[p]$ the volume of product p imported (product units/time) and $T_e[p]$ the volume of the product exported (product units/time). Obviously, S_n can have a negative sign, which means a net water loss instead of a saving.

The global water saving S_g (volume/time) through the trade in product p from an exporting nation n_e to an importing nation n_i, is:

$$S_g[n_e, n_i, p] = T[n_e, n_i, p] \times (WF_{prod}[n_i, p] - WF_{prod}[n_e, p]) \quad \text{[volume/time]} \tag{41}$$

where T is the volume of trade in p (products units/time) between the two nations. The global saving is thus obtained as the difference between the water productivities of the trading partners. When, in a particular case, the importing nation is not able to produce the product domestically, we recommend taking the difference between the global average water footprint of the product and the water footprint in the exporting nation.

The total global water saving can be obtained by summing up the global savings of all international trade flows. By definition, the total global water saving is equal to the sum of the national savings of all nations.

3.7.5 National water dependency versus water self-sufficiency

We define the virtual-water import dependency (*WD*, %) of a nation as the ratio of the external to the total water footprint of national consumption:

$$WD = \frac{WF_{cons,nat,ext}}{WF_{cons,nat}} \times 100 \quad [\%] \tag{42}$$

National water self-sufficiency (*WSS*, %) is defined as the internal divided by the total water footprint of national consumption:

$$WSS = \frac{WF_{cons,nat,int}}{WF_{cons,nat}} \times 100 \quad [\%] \tag{43}$$

Both water dependency and water self-sufficiency can best be calculated on an annual basis or as an average over a period of years.

Self-sufficiency is 100 per cent when all the water needed is available and indeed taken from within the nation's own territory. Water self-sufficiency approaches zero if the demands of goods and services in a nation are heavily met with gross virtual-water imports, in other words, the nation has a relatively large external water footprint in comparison to its internal water footprint.

3.8 Water footprint accounting for catchments and river basins

Full catchment or river basin water footprint accounts are similar to full national water footprint accounts as discussed in the previous section. The only difference is in the definition of the area boundaries considered. National water footprint accounts consider the water footprint within the national territory and the water footprint of the consumers living in that territory. Catchment water footprint accounts combine accounts on the 'water footprint of consumers living in a catchment' (consumer accounts as introduced in Section 3.5) and accounts on the 'water footprint within a catchment area' (area accounts as introduced in Section 3.6). Figure 3.10 shows a visual representation of the catchment water footprint accounting scheme, which indeed looks similar to the national water footprint accounting scheme.

As a guide, one can follow the same method as for national water footprint accounting (Section 3.7). Simply replace 'national' with 'catchment'. The only (practical) difference with national water footprint accounting is the fact that trade data are not available as in the case of nations. One can thus not make use of trade statistics. Instead, trade flows need to be inferred from available data or estimates on production and consumption within the catchment. One can assume that a production surplus (when production > consumption in the catchment) has been exported out of the basin (supposed that there is no storage within the catchment taken to next year); similarly, one can assume that a production shortage (production < consumption) has been imported.

It should be noted that one does not always need to make *full* catchment water footprint accounts as shown in Figure 3.10. It depends on the goal of the accounts. Particularly, catchment managers will primarily be interested in

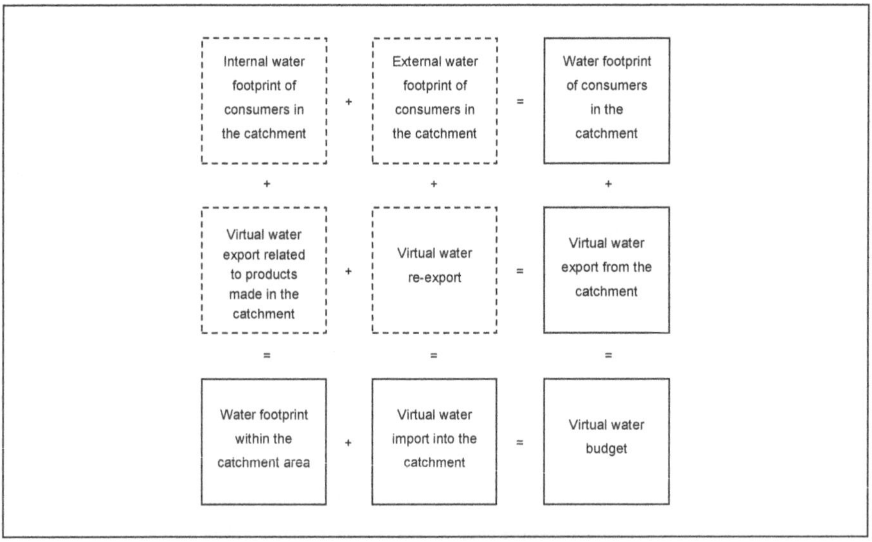

Figure 3.10 The catchment water footprint accounting scheme. The accounting scheme shows the various balances that hold for the water footprint of consumers living within the catchment, the water footprint within the catchment area, the total virtual-water export from the catchment and the total virtual-water import into the catchment

the water footprint within their catchment, not so much in the external water footprint of the people living in the catchment. They may also care little about whether the water footprint within the catchment is for making products consumed by people living in the catchment or for making products that are exported from the catchment. In such a case, one can suffice in making area water footprint accounts as discussed in Section 3.6. However, for a broader understanding of the relation between water use in a catchment and the sustenance of the community living in the catchment, one will need to make full catchment water footprint accounts.

3.9 Water footprint accounting for municipalities, provinces or other administrative units

Water footprint accounts for a municipality, province or other administrative unit resemble water footprint accounts for a nation (Section 3.7) or a catchment (Section 3.8). The same water footprint accounting scheme can be applied (Figures 3.9–3.10). Water footprint accounts at state/province level have been carried out for China (Ma et al, 2006), India (Verma et al, 2009), Indonesia (Bulsink et al, 2010) and Spain (Garrido et al, 2010). At the time of writing, no municipal water footprint studies have been carried out yet. One can expect that the smaller the administrative unit, the larger the external fraction of the water footprint of the consumers in the area will be, most in particular for urban areas.

3.10 Water footprint of a business

3.10.1 Definition

The water footprint of a business is defined as the total volume of freshwater that is used directly or indirectly to run and support the business. It consists of two main components. The operational (or direct) water footprint of a business is the volume of freshwater consumed or polluted due to the business's own operations. The supply chain (or indirect) water footprint of a business is the volume of freshwater consumed or polluted to produce all the goods and services that form the inputs of production of the business. Instead of the term 'business water footprint' one can also use the terms 'corporate water footprint' or 'organizational water footprint'.

The total water footprint of a business can be schematized into components as shown in Figure 3.11. After the distinction between operational and supply chain water footprint, one can differentiate between the water footprint that can

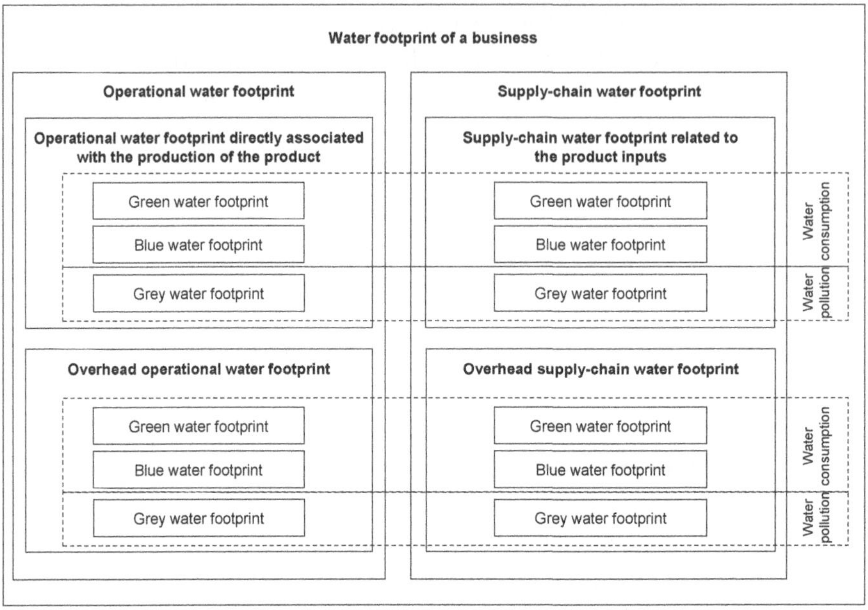

Figure 3.11 Composition of the water footprint of a business

be immediately associated with the product(s) produced by the businesses and the 'overhead water footprint'. The latter is defined as the water footprint pertaining to the general activities for running a business and to the general goods and services consumed by the business. The term 'overhead water footprint' is used

Table 3.1 *Examples of the components of a business water footprint*

Operational water footprint		*Supply chain water footprint*	
Water footprint directly associated with the production of the business's product(s)	*Overhead water footprint*	*Water footprint directly associated with the production of the business product(s)*	*Overhead water footprint*
• Water incorporated into the product. • Water consumed or polluted through a washing process. • Water thermally polluted through use for cooling.	• Water consumption or pollution related to water use in kitchens, toilets, cleaning, gardening, or washing working clothes.	• Water footprint of product ingredients bought by the company. • Water footprint of other items bought by the company for processing their product.	• Water footprint of infrastructure (construction materials and so on). • Water footprint of materials and energy for general use (office materials, cars and trucks, fuels, electricity and so on).

to identify water consumption that is necessary for the continued functioning of the business but that does not directly relate to the production of one particular product. In every case, one can distinguish a green, blue and grey water footprint component. Examples of the various components in a business water footprint are given in Table 3.1.

In addition to the operational and supply chain water footprint, a business may like to distinguish an 'end-use water footprint' of its product. This water footprint refers to the water consumption and pollution by consumers when using the product, for example, think about the water pollution that results from the use of soaps in the household. The end-use water footprint of a product is strictly spoken not part of the business water footprint or the product water footprint, but part of the consumer's water footprint. Consumers can use products in various ways, so that estimating the 'end-use water footprint' of a product will require assumptions about average usage.

By definition, the 'water footprint of a business' is equal to the 'sum of the water footprints of the business output products'. The 'supply chain water footprint of a business' is equal to the 'sum of the water footprints of the business input products'. Calculating a business water footprint or calculating the water footprint of the major product(s) produced by a business is about the same thing, but the focus is different. In the calculation of a business water footprint, there is a strong focus on making the distinction between an operational (direct) and supply-chain (indirect) water footprint. This is highly relevant from a policy perspective, because a business has direct control over its operational water footprint and indirect influence on its supply chain water footprint. When calculating a product water footprint, there is no distinction between the direct and indirect water footprints; one simply considers the process water footprints for all relevant processes within the production system, ignoring how the production system may be owned and operated by different companies. An hybrid between a product and business water footprint account is possible by focussing on the calculation of the water footprint of a particular product – for example, by looking at just one of many products produced by a business – but making explicit which part of the product's water footprint occurs in the business's own operations and which part in the business's supply chain.

Business water footprint accounting offers a new perspective for developing a well-informed corporate water strategy. This is because the water footprint as an indicator of water use differs from the indicator 'water withdrawal in the business's own operations' used by most companies thus far. Box 3.10 discusses a few possible implications for companies that start to look at their water footprint.

Box 3.10 *What is new for companies when considering their business water footprint?*

- Companies have traditionally focused on water use in their operations, not in their supply chain. The water footprint does take an integrated approach. Most companies will discover that their supply chain water footprint is much larger than their operational water footprint. As a result, companies may conclude that it is more cost-effective to shift investments from efforts to reduce their operational water use to efforts to reduce their supply chain water footprint and associated risks.
- Companies have traditionally looked at reduction of water withdrawals. The water footprint shows water use in terms of consumption rather than in terms of withdrawal. Return flows can be reused, so it makes sense to specifically look at consumptive water use.
- Companies make sure that they have a water use right or licence. Possessing that is not sufficient to manage water-related risks. It is useful to look into the spatiotemporal details of a company's water footprint, because details on where and when water is used can be used as input to a detailed water footprint sustainability assessment, to identify the environmental, social and economic impacts and to find out associated business risks.
- Companies have traditionally looked at meeting emission standards (effluent standards). The grey water footprint looks at the required water volume for assimilating waste based on ambient water quality standards. Meeting emission standards is one thing, but looking at how effluents actually result in reduced assimilation capacity of ambient freshwater bodies and at business risks associated to that is another. Meeting effluent standards (which are formulated in terms of concentrations) can easily be done by taking in more water in order to dilute the effluent before disposal. Diluting effluents may be helpful in meeting effluent standards, but not in reducing the grey water footprint, because the latter is related to the total load of chemicals added to the environment, not the concentration of chemicals in the effluent. (This is nicely illustrated in the first example in Appendix IV.)

3.10.2 Choosing the organizational boundaries of the business

A business is conceived here as a coherent entity producing goods and/or services that are supplied to consumers or other businesses. It can be a private company or corporation, but also a governmental or non-governmental organization. It can refer to various levels of scale, for instance, a specific unit or division of a company, an entire company or a whole business sector. In the public sector, one may refer to a unit within a municipality as well as to national government as a whole. The term business can also refer to a consortium or joint venture of companies or organizations aimed at the delivery of a certain good or service. In fact, the term business can also refer to any project (such as the construction of a piece of infrastructure) or activity (for example, the organization of a large sports event). In this way, the term 'business' has been defined so broadly that it can refer to all sorts of corporations, organizations, projects and activities. In technical terms, a business is understood here as any coherent entity or activity that transforms a set of inputs into one or more outputs.

In order to be able to assess the water footprint of a business, the business should be clearly delineated. It should be clear what are the boundaries of the business considered. It should be possible to schematize the business into a system that is clearly distinguished from its environment and where inputs and outputs are well-known.

Whatever the type of company, companies often consist of a number of units. For example, a company can have operations (such as factories) at various locations. Or a company may have separate divisions at one location. For the purpose of water footprint accounting, it is often useful to distinguish between different business units. For instance, when a manufacturing company has different factories at different locations, the individual factories are likely to operate under different conditions and derive their inputs from different places. In such a case, it is useful to do water footprint accounting per business unit first and aggregate the business unit accounts later on into an account for the business as a whole.

The business needs to be defined by describing the business units that will be distinguished and specifying the annual inputs and outputs per business unit. Inputs and outputs are described in physical units. Preferably, a business unit refers to a part of the total business that produces one particular product at one particular spot. When a business runs at different locations, it is thus preferred to schematize the overall business into business units in such a way that individual business units operate at one location. Besides, operations of a business at one particular spot are preferably schematized in different business units each producing its own product. It is most useful to schematize the business based on the various primary products delivered by the business. However, one can also distinguish service units providing only goods or services to primary production units.

As an example, Figure 3.12 shows a business producing output products A, B and C. The business consists of three business units. Unit 1 produces product A. Part of A is delivered to business unit 2, but most is sold to other businesses.

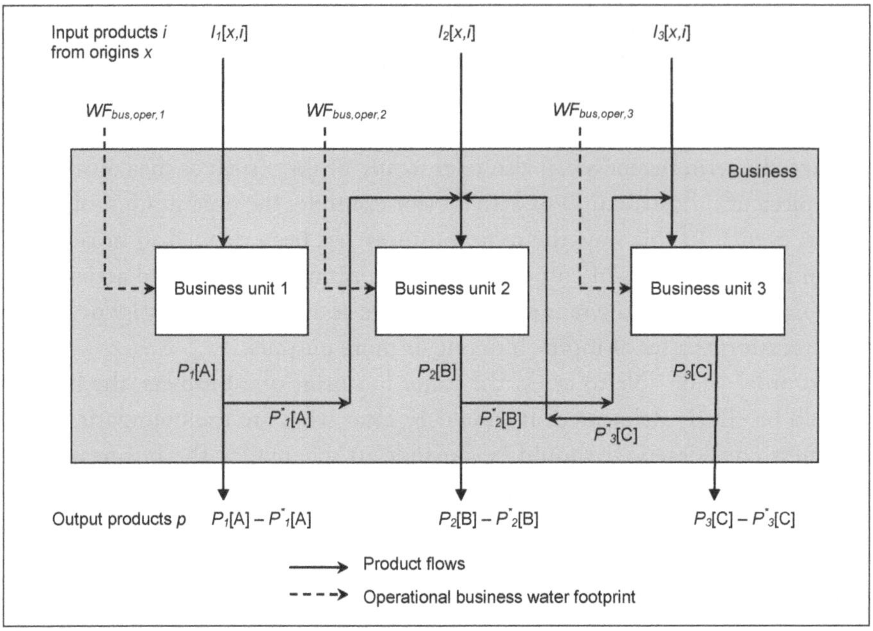

Figure 3.12 Business that consists of three business units producing products A–C. Product inflow $I_u[x,i]$ refers to the annual volume of input product i from source x into business unit u. Product outflow $P_u[p]$ refers to the annual volume of output product p from business unit u. Product flow $P^*_u[p]$ refers to the part of $P_u[p]$ that goes to another business unit within the same business.

Unit 2 produces product B, which is partly sold to another business and partly delivered to unit 3. Unit 3 produces product C, both for delivery to unit 2 and for selling externally. Each unit has an intake of a number of input products derived from companies in a preceding link of the production chain, and a related indirect freshwater input, as well a direct freshwater input. A schema like the one shown in Figure 3.12 can form the basis for calculating a business water footprint, as will be explained in the next section.

When a business is large and heterogeneous (different locations, different products), it can be attractive to schematize the business into some major business units and each major unit into a number of minor units again. In this way the business can be schematized as a system with subsystems at a number of levels. Later on the water footprint accounts at the lowest level can be aggregated to accounts at the second-lowest level, and so on, up to the level of the business as a whole.

3.10.3 Calculation of the business water footprint

Below we will show how one can calculate the water footprint of a 'business unit'. At the end of the section, it will be shown how one can calculate the water footprint of a business consisting of a number of business units. The water footprint of a business unit (WF_{bus}, volume/time) is calculated by adding the operational water footprint of the business unit and its supply-chain water footprint:

$$WF_{bus} = WF_{bus,oper} + WF_{bus,sup} \quad \text{[volume/time]} \tag{44}$$

Both components consist of a water footprint that can be directly associated with the production of the product in the business unit and an overhead water footprint:

$$WF_{bus,oper} = WF_{bus,oper,inputs} + WF_{bus,oper,overhead} \quad \text{[volume/time]} \tag{45}$$

$$WF_{bus,sup} = WF_{bus,sup,inputs} + WF_{bus,sup,overhead} \quad \text{[volume/time]} \tag{46}$$

The operational water footprint is equal to the consumptive water use and the water pollution that can be associated with the operations of the business. Following the guidelines provided in Section 3.3, one can simply look at the evaporative flow from the operations, the volume of water incorporated into products and the return flows of water to other catchments than from which water was withdrawn. In addition, one has to consider effluent volumes and concentrations of chemicals therein. The operational overhead water footprint – water consumption and pollution related to general water-using activities in the business unit – can be identified and quantified just like the operational water footprint directly associated with the production process. The overhead water footprint, however, will often serve more than the business unit considered. For example, the overhead of a factory with two production lines will have to be distributed over the two production lines. If one has defined a business unit such that it refers to one of the production lines, one needs to calculate the share of the overhead water footprint that is to be accounted to the one production line. One can do this according to the production values of the two production lines.

The supply chain water footprint per business unit (volume/time) can be calculated by multiplying the various input-product volumes (data available from the business itself) by their respective product water footprints (data that have to be obtained from suppliers). Supposed that there are different input products i originating from different sources x, the supply-chain water footprint of a business unit is calculated as:

$$WF_{bus,sup} = \sum_x \left(\sum_i \left(WF_{prod}[x,i] \times I[x,i] \right) \right) \quad \text{[volume/time]} \tag{47}$$

in which $WF_{bus,sup}$ represents the supply-chain water footprint of the business unit (volume/time), $WF_{prod}[x,i]$ the water footprint of input product i from source x (volume/unit of product) and $I[x,i]$ the volume of input product i from source x into the business unit (product units/time).

The product water footprint depends on the source of the product. When the product comes from another business unit within the same business, the value of the product water footprint is known from the business's own accounting system (see the end of this section). When the product originates from a supplier outside the business, the value of the product water footprint has to be obtained from the supplier or estimated based on indirect data known about the production characteristics of the supplier. The various product water footprints are composed of three colours (green, blue, grey), which should be accounted separately, so that the resulting supply chain water footprint of the business unit consists of three colour-components as well.

The water footprint of each specific output product of a business unit is estimated by dividing the business-unit water footprint by the output volume. Allocation of the water footprint over the output products can be done in several ways, for example, according to mass, energy content or economic value. Following what is common in life cycle assessment studies, it is recommended to allocate according to economic value. The product water footprint of output product p from a business unit ($WF_{prod}[p]$, volume/unit of product) can then be calculated as:

$$WF_{prod}[p] = \frac{E[p]}{\sum_p E[p]} \times \frac{WF_{bus}}{P[p]} \quad \text{[volume/product unit]} \tag{48}$$

in which $P[p]$ is the volume of output product p from the business unit (product units/time), $E[p]$ the total economic value of output product p (monetary unit/time) and $\sum E[p]$ the total economic value of all output products together (monetary unit/time). If the business unit delivers only one product, the equation is reduced to:

$$WF_{prod}[p] = \frac{WF_{bus}}{P[p]} \quad \text{[volume/product unit]} \tag{49}$$

All above equations are to be applied at the level of a business unit. Suppose that a business has been schematized into a number of business units u, the water footprint of the business as a whole ($WF_{bus,tot}$) is calculated by aggregating the water footprints of its business units. In order to avoid double counting, one has to subtract the virtual-water flows between the various business units within the business:

$$WF_{bus,tot} = \sum_{u} WF_{bus}[u] - \sum_{u}\sum_{p}\left(WF_{prod}[u,p] \times P^{*}[u,p]\right) \quad \text{[volume/time]} \qquad (50)$$

in which $P^{*}[u,p]$ stands for the annual volume of output product p from business unit u to another business unit within the same business (product units/time).

Notes

1 It is recognized that water use connected to a product is not limited to its production stage. In the case of many products (for example, a washing machine), there is some form of water use involved in the use stage of the product. This component of water use, however, is not part of the product water footprint. The water use during product use is included in the water footprint of the consumer of the product. Water use in the reuse, recycle or disposal stage of a product is included in the water footprint of the business or organization that provides that service and is included in the water footprints of the consumers that benefit from that service.

2 This assumption implies that $\frac{WF_{cons,nat,ext}}{V_{e,r}} = \frac{WF_{cons,nat,int}}{V_{e,d}} = \frac{WF_{cons,nat}}{V_{e}}$ and that $\frac{WF_{cons,nat,ext}}{WF_{cons,nat,int}} = \frac{V_{e,r}}{V_{e,d}} = \frac{V_{i}}{WF_{area,nat}}$.

3 One should make an exception for product categories for which re-export is a substantial part of import. The national ratio between $WF_{cons,nat,ext}$ and V_i is not a good assumption here. Instead, one could apply a specific ratio of $WF_{cons,nat,ext}$ to V_i valid to the product category considered.

Chapter 4

Water Footprint Sustainability Assessment

4.1 Introduction

The water footprint is an indicator of freshwater appropriation (in m^3/yr), developed as an analogue to the ecological footprint, which is an indicator of use of biologically productive space (in ha). In order to get an idea of what the footprint size means, one will need to compare the water footprint to the available freshwater resources (also expressed in m^3/yr), in the same way as one needs to compare the ecological footprint to the available biologically productive space (in ha) (Hoekstra, 2009). In essence, water footprint sustainability assessment is primarily about making this comparison of the human water footprint with what the Earth can sustainably support. When diving into this issue, however, one will discover that there are many different sorts of questions that one can pose and that there are many complexities involved. Sustainability, for instance, has different dimensions (environmental, social, economic), impacts can be formulated at different levels (primary, secondary impacts) and the water footprint has different colours (green, blue, grey). In this chapter we present a guide to water footprint sustainability assessment, which has grown out of the recent past of increased attention to the subject (Box 4.1).

The question of the sustainability of water footprints can be considered from a number of distinct points of view. From the geographical viewpoint, one can ask: Is the total water footprint within a certain geographic area sustainable? This will not be the case when, for example, environmental flow requirements or ambient water quality standards in a catchment area are compromised or when the water allocation within a catchment area is regarded as unfair or inefficient. When considering a specific water-using process, one can ask: Is the water footprint of this process sustainable? The answer depends on two criteria. First, the water footprint of a process is unsustainable when the process is situated in a certain period of the year in a certain catchment or river basin in which the overall water footprint is unsustainable. Second, the water footprint of a process is unsustainable in itself – independent of the geographic context – when either the green, blue or grey water footprint of the process can be reduced or avoided altogether (at acceptable societal cost). From the viewpoint of a product, the

Box 4.1 *History of water footprint sustainability assessment*

During the first years of the water footprint concept (2002–2008), the focus was on water footprint accounting. The water footprint is primarily an innovation regarding how to measure freshwater appropriation by human beings. Before, water was not measured along supply chains and green and grey water were excluded from water use statistics. Besides, measurements focused on blue water withdrawals, ignoring the fact that it is particularly consumptive use that determines the impact on the water system of a catchment. From the beginning of the concept it was recognized that water footprint accounting is about freshwater appropriation only, that the green and blue water footprints in a geographic area need to be compared to the green and blue water that is available and that the grey water footprint needs to be compared to the capacity of a catchment to assimilate the waste. Little work on that was done, however. Hoekstra (2008a) made for the first time explicit mentioning of the need for a 'sustainability assessment' phase after the accounting phase, although at that time it was called 'impact assessment'. Comparing water footprints to actual water availability and identifying hotspots of scarcity was done for the first time by Van Oel et al (2008), Kampman et al (2008) and Chapagain and Orr (2008).

In the first *Water Footprint Manual*, the term 'impact assessment' was changed into 'sustainability assessment' because the latter term better reflects what it should encompass (Hoekstra et al, 2009a). The term 'impact' draws the attention to local impacts, namely the immediately visible impacts on the ground, which is a too restricted view. Since freshwater resources in the world are limited, one needs to look at the sustainability of water footprints in a much broader context. Overlooking the issue of water use or pollution in water-abundant areas is similar to ignoring energy use in oil-rich countries. Wasting water or polluting water in a certain water-rich geographic area is a concern just like wasting energy in oil-rich countries. Efficient use of water in water-rich areas to produce water-intensive commodities reduces the need to use water for producing those commodities in water-scarce areas. Therefore, assessing the sustainability of water footprints is more than looking at whether a water footprint has immediate local impacts.

An initial step to give structure to the phase of 'water footprint sustainability assessment' was made in the *Water Footprint Manual*. Subsequently, in the period December 2009–July 2010 the Water Footprint Sustainability Assessment Working Group of the Water Footprint Network (WFN) reflected on the manual and provided a number of recommendations (Zarate, 2010b) that have been taken into the process of preparing this version of the manual. Most notably we have adopted the idea of 'environmental sustainability boundaries' (after Richter, 2010) and the concept of primary-secondary impacts. As one will note when comparing the current manual with the first one, the chapter on sustainability assessment has been completely restructured. It is better recognized now that estimating the sustainability of the total water footprint in a geographic area is something different from estimating the sustainability of a particular process, product, producer or consumer. A clear distinction is made between the different sorts of questions that exist and the different ways they need to be addressed.

relevant question is: Is the water footprint of the product sustainable? The answer to this question will depend on the sustainability of the water footprints of the processes that are part of the production system to make the product. From the viewpoint of the producer, one can ask: Is the water footprint of the producer sustainable? Since the water footprint of a producer is equal to the sum of the water footprints of the products produced by the producer, the answer to this question depends on the sustainability of the products made by the producer. Finally, from the viewpoint of a consumer, one can ask: Is the water footprint of the consumer sustainable? Since the water footprint of a consumer is equal to the sum of the water footprints of the products consumed by the consumer, the answer will depend again on the sustainability of the water footprints of the products consumed. Here, however, an additional criterion comes into play, because the sustainability of the water footprint of a consumer also depends on whether the consumer water footprint is smaller or larger than an individual's fair share given the limitations to the water footprint of humanity.

The sustainability of the water footprint of a product, producer or consumer partly depends on the geographic contexts in which the various water footprint components of a product, producer or consumer are located. It is seldom the water footprint of one particular process, product, producer or consumer that creates the problems of water scarcity and pollution as we experience them. Those problems emerge as the cumulative effect of all activities in the geographic area considered. The total water footprint in an area is the sum of many smaller footprints, each one of which will be related to a certain process, product, producer and consumer. When the water footprint of a process, product, producer or consumer *contributes* to the unsustainable situation observed within a certain geographical context, one can say that this water footprint is unsustainable as well.

In this chapter we shall start by showing how to assess the sustainability of the water footprint within a catchment or river basin. Subsequently, we will show how to assess the sustainability of the water footprint of a process, product, producer and consumer. The order has been chosen this way, because in the later sections we will refer back to the earlier sections. One cannot assess the sustainability of the water footprint of a process without knowing the sustainability of the total water footprint in the catchment where the process is located. One cannot assess the sustainability of the water footprint of a product without knowing the sustainability of the processes involved. And, finally, one cannot assess the sustainability of the water footprints of producers or consumers without knowing the sustainability of the products being produced or consumed.

4.2 Geographic sustainability: Sustainability of the water footprint within a catchment or river basin

4.2.1 Introduction

Evaluating the sustainability of the total water footprint in a geographic area can best be done at the level of a catchment area or river basin. At the level of such a hydrological unit, one can reasonably compare the green or blue water footprint to the green or blue water availability, or the grey water footprint to the available waste assimilation capacity. Also issues of fair and efficient water resources allocation are most relevant on the scale of a catchment or river basin.

The sustainability of the water footprint within a catchment or river basin can be analysed from three different perspectives: environmental, social and economic. From each of the perspectives there are some 'sustainability criteria' (Box 4.2). A sustainability criterion spells out when a water footprint in a catchment or basin can no longer be considered sustainable.

Identifying and quantifying the sustainability criteria is the first step when assessing the sustainability of the water footprint in a catchment or river basin (Figure 4.1). The second step is to identify hotspots within the catchment or river basin, namely sub-catchments and periods within the year in which the water footprint is regarded as unsustainable. In the third and fourth steps, one will quantify the primary and secondary impacts in the hotspots.

A hotspot is a specific period of the year (for example, the dry period) in a specific (sub)catchment in which the water footprint is unsustainable, perhaps because it compromises environmental water needs or water quality standards or because the water allocation and use in the catchment is considered unfair and/or economically inefficient. In a hotspot, problems of water scarcity, pollution or conflict occur. Hotspots are the places where and periods within the year when water footprints are not sustainable and thus have to be reduced.

When one considers a catchment or river basin as a whole, without considering sub-catchments, the catchment or river basin as a whole can be categorized as a

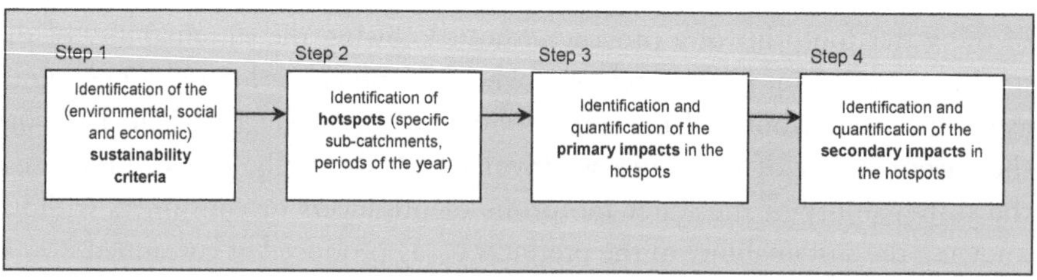

Figure 4.1 Assessment of the sustainability of the water footprint within a catchment or river basin in four steps

Box 4.2 *Sustainability criteria for water use and allocation within a catchment or river basin*

The water footprint within a catchment needs to meet certain criteria in order to be sustainable. Sustainability has an environmental dimension as well as a social and economic dimension.

- **Environmental sustainability:** Water quality should remain within certain limits. As an indicator of what these limits are, one can best consider 'ambient water quality standards' that people have agreed upon. In addition, river and groundwater flows should remain within certain limits compared to natural run-off, in order to maintain river and groundwater-dependent ecosystems and the livelihoods of the people that depend on those ecosystems. In the case of rivers, the so-called 'environmental flow requirements' form boundaries for run-off alteration, comparable to the way in which water quality standards form boundaries for pollution (Richter, 2010). In the case of green water, the 'environmental green water requirements' form the boundaries for green water appropriation for human purposes.
- **Social sustainability:** A minimum amount of the freshwater available on Earth needs to be allocated to 'basic human needs', most notably a minimum domestic water supply for drinking, washing and cooking and a minimum allocation of water to food production to secure a sufficient level of food supply to all. This criterion implies that only the fraction of available freshwater supply that remains after subtraction of environmental water needs and water requirements to sustain basic human needs can be allocated to 'luxury' goods. A minimum domestic water supply for drinking, washing and cooking needs to be guaranteed at the catchment or river basin level. A minimum allocation of water to food production is to be secured at global level, since river basin communities are not necessarily self-sufficient in food, provided that food security is ensured through food imports.
- **Economic sustainability:** Water needs to be allocated and used in an economically efficient way. The benefits of a (green, blue or grey) water footprint that results from using water for a certain purpose should outweigh the full cost associated with this water footprint, including externalities, opportunity costs and a scarcity rent. If this is not the case, the water footprint is unsustainable.

When the green, blue or grey water footprint in a catchment does not fulfil one of the criteria of environmental, social or economic sustainability, the water footprint cannot be considered as 'geographically sustainable'.

hotspot area or not, depending on whether problems emerge at the level of the catchment or basin as a whole. The advantage of looking for hotspots at the level of relatively small catchment areas (for example, up to 100 km^2) is that one may be able to identify hotspots that would disappear at the course resolution level of larger catchments or a river basin as a whole. If one compares the grey water footprint within a basin with the waste assimilation capacity in the basin as a whole, it may show that there is enough waste assimilation capacity, while this may not be the case in some specific upstream sub-catchments where most of the pollution in the basin is concentrated. The disadvantage of looking for hotspots at a fine spatial resolution is that much more data are needed (on how the total green, blue and grey water footprints in a basin and on how green and blue water availability and waste assimilation capacity are spatially distributed within the basin). Another disadvantage is that some problems may emerge only at a larger spatial scale, for example because pollutants accumulate downstream. The best approach is therefore to take a whole river basin as the analytical unit and to distinguish relatively small sub-catchments within this basin. The assessment can then be done at both the fine level of sub-catchments and the aggregate level of larger catchments and the river basin as a whole.

After having identified hotspots, one can study the environmental, social and economic implications in more detail. We distinguish between identifying 'primary' and 'secondary' impacts. Primary impacts are described in terms of changed water flows and quality (compared to the natural conditions, without human disturbances). One will show, for example, how much run-off in a catchment has decreased by man's blue water footprint in the catchment and to which extent this conflicts with environmental flow requirements. Or one describes in detail how water quality has changed compared to natural conditions, for instance per water quality parameter, and which parameters violate the ambient water quality standards. Secondary impacts are the ecological, social and economic goods or services that are impaired in a catchment area as a result of the primary impacts. Secondary impacts can be measured for instance in terms of lost species, reduced biodiversity, reduced food security, affected human health, reduced income from water-dependent economic activities, and so on.

4.2.2 Environmental sustainability criteria for identifying environmental hotspots

The water footprint in a catchment is environmentally unsustainable and thus creates an environmental hotspot when environmental water needs are violated or when pollution exceeds waste assimilation capacity. In order to have an indication of the severity of a hotspot, one can calculate the green water scarcity, blue water scarcity and water pollution level as will be defined below.

We speak about an environmental hotspot when the green water scarcity, blue water scarcity and/or the water pollution level exceeds 100 per cent. In the case of the blue water footprint, it is also important to evaluate whether the footprint results in a drop in groundwater or lake levels to an extent that these drops exceed a certain environmental threshold.

Environmental hotspots can specifically relate to the green, blue or grey water footprint in a catchment, so they will be discussed in the following one by one. What follows can be applied for catchments of various sizes, including river basins as a whole.

Environmental sustainability of the green water footprint

Whether a total green water footprint in a catchment is actually significant or not will become clear when it is put in the context of how much green water is available. A green water footprint in a specific catchment forms an environmental hotspot when it exceeds the availability of green water. The 'green water availability' (WA_{green}) in a catchment x in a certain period t is defined as the total evapotranspiration of rainwater from land (ET_{green}) minus the evapotranspiration from land reserved for natural vegetation (ET_{env}) and minus the evapotranspiration from land that cannot be made productive:

$$WA_{green}[x,t] = ET_{green}[x,t] - ET_{env}[x,t] - ET_{unprod}[x,t] \quad \text{[volume/time]} \qquad (51)$$

All variables are expressed here in terms of volume/time. The variable ET_{env} is the 'environmental green water requirement' and refers to the quantity of green water used by natural vegetation in areas within the catchment that are to be reserved for nature, in order to preserve biodiversity and support human livelihoods that depend on the natural ecosystems. The environmental green water requirement can be quantified by looking at the evapotranspiration from the land areas that need to be protected from a nature conservation point of view (Box 4.3). The variable ET_{unprod} refers to evapotranspiration that cannot be made productive in crop production, in other words, evapotranspiration in areas or periods of the year that are unsuitable for crop growth. Think of evapotranspiration from mountainous areas where slopes do not allow crop cultivation, evaporation from built-up areas or evapotranspiration in periods that are too cold for crop cultivation (in the latter cases ET is generally low, so this unproductive flow is not so large).

In order to better understand the concept of green water availability, let us give an example. Consider a catchment area of 1000 km^2. On average, annual evapotranspiration is 450 mm, so that total evapotranspiration (ET_{green}) from the catchment is 1000 km^2 × 450 mm = 450 million m^3. Presume that studies have shown that for biodiversity preservation 30% of the catchment area needs to be reserved for nature and that the evapotranspiration from this area over a year is 500 mm on average. The environmental green water requirement (ET_{env}) in the catchment is thus 0.3 × 1000 km^2 × 500 mm = 150 million m^3. Further suppose that another 30% of the catchment area is not suitable for crop growth (built-up areas including roads and other infrastructure) and that average annual evapotranspiration in these areas is 400 mm. The evapotranspiration from this unproductive area is 0.3 × 1000 km^2 × 400 mm = 120 million m^3. The annual evapotranspiration in the remainder of the catchment is 0.4 × 1000 km^2 × 450 mm = 180 million m^3. Half of the year, during the winter, climate is not suitable for crop production, but during this half of the year total evapotranspiration is relatively low, 100 mm. Evapotranspiration in this unproductive period in the area available for farming is thus 0.4 × 1000 km^2 × 100 mm = 40 million m^3. The total evapotranspiration in the catchment that cannot be made productive in crop production (ET_{unprod}) is thus 120+40=160 million m^3. From this example one can see that, although total evapotranspiration (ET_{green}) from the catchment is 450 million m^3, the green water availability is only 450 – 150 – 160 = 140 million m^3.

When people speak about water-scarcity, they generally refer to blue water scarcity. However, green water availability is also limited, so green water resources are scarce as well. The level of green water scarcity in a catchment *x* in a period *t* is defined as the ratio of the total of green water footprints in the catchment to the green water availability.

$$WS_{green}[x,t] = \frac{\sum WF_{green}[x,t]}{WA_{green}[x,t]} \quad [-] \qquad (52)$$

Defined in this way, the green water scarcity indicator in fact denotes the 'fraction of appropriation' of available green water resources. Measuring green water scarcity can be done at a daily basis, but a monthly basis will generally be sufficient to see the variation within the year. A green water scarcity of 100 per cent means that the available green water has been fully consumed. Scarcity values beyond 100 per cent are not sustainable.

It must be admitted here that the issue of analysing green water scarcity is largely unexplored. The difficulty lies in the estimation of 'green water availability'. Particularly, data are lacking on the environmental green water requirement (Box 4.3) and on the quantities of evapotranspiration that cannot be made productive in crop production. These quantities severely limit green water availability and are therefore vital to take into consideration, but without consensus regarding how much land and associated evapotranspiration are to be reserved for nature and about how to precisely define when and where evapotranspiration cannot be made productive, it is impossible to proceed with

Box 4.3 *Environmental green water requirement*

A significant part of the evaporative flow from land needs to be reserved for natural vegetation. When land is conserved for nature, automatically the evaporative flow from that land has been reserved for nature and is no longer available for producing crops or trees for human use. When we want to know how much of the green water availability needs to be subtracted from the total evapotranspiration from land in order to know how much green water is left for human use, we can look at how much and which lands are considered as lands to be reserved for nature. In their Global Strategy for Plant Conservation (CBD, 2002), the Convention on Biological Diversity formulated targets for the year 2010: at least 10 per cent of each of the world's ecological regions should be effectively conserved and 50 per cent of the most important areas for plant diversity should be protected. Estimates for how much of our global land area needs to be reserved for nature range widely. According to the World Commission on Environment and Development (WCED, 1987), at least 12 per cent of all ecosystem types will need to be preserved for biodiversity protection. Noss and Cooperrider (1994) estimate that most regions will need to reserve 25–75 per cent of the land for securing biodiversity. Svancara et al (2005) compared more than 200 targets with respect to the percentage of land to be conserved for biodiversity protection, as proposed in different reports, and concluded that the average percentages of area recommended for evidence-based targets were nearly three times as high as those recommended in policy-driven approaches. Reserving 10–15 per cent of the total land for biodiversity protection as proposed in policy-driven processes (such as CBD, 2002; WCED, 1987) would not represent the actual biological need, which would rather be in the order of 25–50 per cent. The percentage should vary across regions depending on the characteristics of the region. When data for a specific catchment are lacking we recommend to reckon with a default value of at least 12 per cent. More realistic from an ecological point of view is probably to work with a default value of 30 per cent.

a quantitative analysis. This is obviously a field for further research. For the time being, we recommend excluding quantitative assessment of green water scarcity in practical policy settings, but including it in pilot studies to explore the usefulness of such analysis and to work on an unambiguous definition of green water availability.

It is noted here that – when there is a difference between the evapotranspiration from natural vegetation and a crop field – the green water footprint may affect blue water availability, although the effect on river basin scale will generally be small. Generally, this effect can therefore be neglected (Box 4.4).

Box 4.4 *The effect of the green water footprint on blue water availability*

The green water footprint in a catchment can result in a changed run-off pattern downstream. Often, evapotranspiration of rainfall from a crop field will not differ much from evapotranspiration from the field under natural conditions, but it may differ significantly during particular parts of the year. At times evapotranspiration may be lower, at other times higher, leading to increased or reduced run-off respectively. This means that a *green* water footprint can affect *blue* water availability. It has been suggested to speak about the 'net green water footprint' to refer to the difference between the evapotranspiration from the crop and the evapotranspiration under natural conditions (SABMiller and WWF-UK, 2009). This terminology, however, is not consistent with the basic definition of the water footprint concept as an indicator of freshwater appropriation, which requires that we look at totals. We recommend speaking about 'changed run-off as a result of the green water footprint' instead of 'net green water footprint'. Agriculture is not the only human factor that affects the fraction of the precipitation that becomes run-off (thus affecting blue water availability); also factors such as urbanization and other landscape changes will affect run-off and therefore blue water availability. These issues are to be considered when assessing blue water scarcity and its underlying causes. Since blue water scarcity refers to the ratio between the blue water footprint in an area and the blue water availability, blue water scarcity can be increased either by an increased blue water footprint or by decreased blue water availability. In all river basins the historical increase in the blue water footprint has been so much bigger than changes in blue water availability, that when considering blue water scarcity it may suffice to consider the changing blue water footprint against the background of the blue water availability as a geographical constant.

Environmental sustainability of the blue water footprint

The total blue water footprint in a catchment is equal to the aggregate of all blue process water footprints within the catchment. A blue water footprint in a specific period in a specific catchment forms a hotspot when the blue water footprint exceeds blue water availability. The blue water availability (WA_{blue}) in a catchment x in a certain period t is defined as the natural run-off in the catchment (R_{nat}) minus the so-called 'environmental flow requirement' (EFR):

$$WA_{blue}[x,t] = R_{nat}[x,t] - EFR[x,t] \quad \text{[volume/time]} \qquad (53)$$

When, in a certain period and catchment, the blue water footprint exceeds the blue water availability, this means that the environmental flow requirement in that period and catchment is violated. Environmental flow requirements are

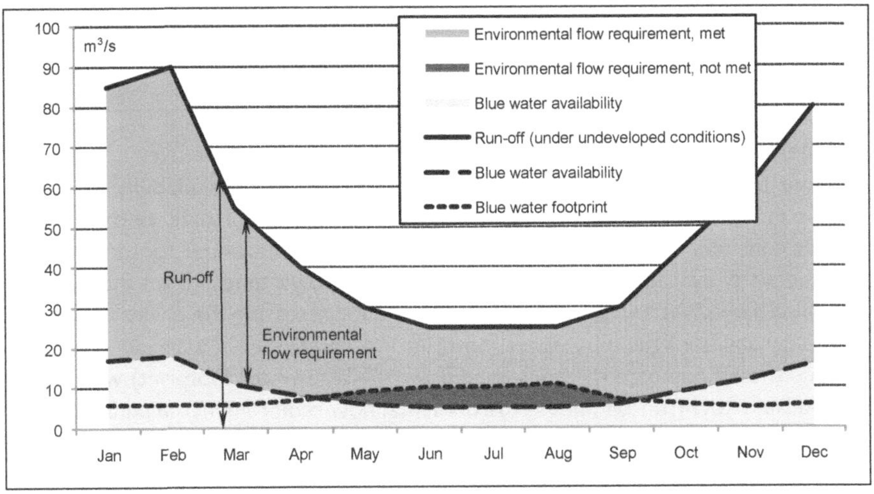

Figure 4.2 The blue water footprint over a year compared to the blue water availability, where the latter is equal to run-off (under undeveloped conditions) minus environmental flow requirements

to be formulated in terms of the quantity and timing of water flows required to sustain freshwater and estuarine ecosystems and the human livelihoods and well-being that depend on these ecosystems. Appendix V discusses the concept of environmental flow requirement in more detail. Figure 4.2 shows how the blue water footprint over the year can be compared with the blue water availability over the year. In the case shown, environmental flow requirements are violated during a certain period of the year, but not during the rest of the year. Environmental flow requirements are subtracted from the natural run-off, not the actual run-off, because the latter has already been affected by upstream water consumption. The natural run-off can be estimated as the actual run-off plus the blue water footprint within the catchment.

When, in a certain month, the blue water footprint within a catchment exceeds the blue water availability, the blue water footprint is environmentally unsustainable, because environmental flow requirements are violated. There are, however, more criteria that have to be considered. The blue water footprint in a catchment will not only affect the run-off flow, but it will also affect blue-water stocks in the basin, most notably groundwater stocks and water volumes in lakes. Therefore, an additional way of identifying unsustainable conditions is to look at the effect of the blue water footprint on groundwater and lake levels in the catchment (Box 4.5).

The 'blue water scarcity' in a catchment x (WS_{blue}) is defined as the ratio of the total of blue water footprints in the catchment (ΣWF_{blue}) to the blue water availability (WA_{blue}):

Box 4.5 *The sustainability of a blue water footprint depends on how it affects both blue-water flows and stocks*

The blue water footprint in a catchment, given as a volume over a certain time period, should be compared to the blue water availability in the catchment over the same period. Obviously, per unit of time, one should not consume more than what is available. Both the blue water footprint and blue water availability are expressed as a volume over time and are therefore called 'flows'. As explained in the main text, for assessing the environmental sustainability of the blue water footprint in a catchment in a certain period, one should compare the consumed flow (the blue water footprint) with the available flow (run-off minus environmental flow requirement). In addition to this, however, one should also look at the effect of the blue water footprint on the blue water stocks (the water volumes stored in the ground and in lakes). We can illustrate this with a simple example.

Imagine a lake that is fed by a river at one side and drained by the same river on the other side. For simplicity, we assume that precipitation on and evaporation from the lake are relatively small compared to the river flow, so that the river outflow is equal to the river inflow. Suppose that the river inflow gets reduced by 20 per cent as a result of the total upstream blue water footprint. The lake level will drop until a level has reached at which the outflow from the lake will be equal to the inflow again. The lake has then arrived at a new equilibrium, with a lower water volume and associated lower water level. When the lake outflow is linearly dependent on the active lake volume (the volume above the height of the bottom of the outflow point), then a 20 per cent decrease in outflow corresponds to a 20 per cent decrease in active lake volume, which will result in a certain decline of the lake level. The question, now, is not only whether the 20 per cent *reduction in river flow* is sustainable, but also whether the 20 per cent *decrease in the lake water volume* and corresponding *lake level decline* is sustainable. The former depends on the environmental flow requirements. The latter depends on the 'maximum allowable lake level decline', which depends on the vulnerability of the lake's aquatic ecosystem and the riparian ecosystems to a changing water level. A similar example can be given for renewable groundwater reserves. The net groundwater abstraction from an aquifer should remain below the groundwater availability (the rate of groundwater recharge minus the fraction of natural groundwater outflow required to sustain environmental flow requirements in the river). In addition, however, one should look how groundwater abstractions affect the groundwater table. Whether a certain groundwater level decline is sustainable, depends on the 'maximum allowable groundwater level decline', which depends on the vulnerability of terrestrial ecosystems to a changing groundwater level. A warning has to be made here that when measuring groundwater and lake level decline one should be careful in distinguishing natural and human causes. Intra- and inter-annual variations in groundwater and lake levels are natural, related to climatic variability. Before attributing groundwater or lake level decline to the blue water footprint in the catchment, one should verify that the drop is not due to changes in climatic parameters within the period considered. Similar to environmental flow requirements, maximum allowable lake and groundwater level declines are context dependent and therefore need to be estimated separately per catchment.

In order to assess the sustainability of the blue water footprint in a catchment, it is thus necessary to make the comparison with blue water availability but, in addition, one needs to evaluate whether lake or groundwater levels remain within their 'sustainability boundaries' (Richter, 2010). Fossil groundwater is a case in itself. When the blue water footprint is based on fossil groundwater, every drop of water consumed will subtract from the available fossil groundwater stock. Fossil groundwater consumption is always depletion and thus by definition unsustainable altogether.

$$WS_{blue}[x,t] = \frac{\sum WF_{blue}[x,t]}{WA_{blue}[x,t]} \quad [-] \tag{54}$$

A blue water scarcity of 100 per cent means that the available blue water has been fully consumed. A blue water scarcity beyond 100 per cent is not sustainable. The blue water scarcity is time-dependent; it varies within the year and from year to year. Measuring can be done at daily basis, but a *monthly* basis will generally be sufficient to see the variation within the year. From Figure 4.2 one can see that measuring blue water scarcity on an annual basis is not a good idea. In the example given it is clear that during five months of the year (May–September) the water scarcity is beyond 100 per cent. During the other seven months, the water scarcity is below 100 per cent. Averaging the monthly scarcity values over all months of the year in the given example, gives an average monthly blue water scarcity in this catchment of slightly larger than 100 per cent. If, in the example, one takes the annual blue water footprint over the annual blue water availability, one gets a value of 75 per cent, a figure that obscures the fact that environmental flow requirements are violated during five months of the year, the entire dry period!

It should be noted that the blue water scarcity as defined here is a physical and environmental concept. It is physical because it compares appropriated to available volumes and environmental because it accounts for environmental flow needs. It is not an economic scarcity indicator; an economic scarcity indicator would use monetary values to express scarcity. Further, the blue water scarcity indicator as defined above differs from conventional water scarcity indicators in a number of ways, and basically tries to improve upon the shortcomings of those conventional scarcity indicators (Box 4.6).

Box 4.6 *How the 'blue water scarcity' as defined in water footprint studies differs from conventional water scarcity indicators*

Water-scarcity indicators are always based on two basic ingredients: a measure of water use and a measure of water availability. The most common indicator of blue water scarcity is the ratio of the annual water withdrawal in a certain area to the total annual run-off in that area, called variously the water utilization level (Falkenmark, 1989), the withdrawal-to-availability ratio (Alcamo and Henrichs, 2002) or the use-to-resource ratio (Raskin et al, 1996). There are four critiques to this approach. First, water withdrawal is not the best indicator of water use when one is interested in the effect of the withdrawal at the scale of the catchment as a whole, because water withdrawals partly return to the catchment (Perry, 2007). Therefore it makes more sense to express blue water use in terms of consumptive water use, in other words, by considering the blue water footprint. Second, total run-off is not the best indicator of water availability, because it ignores the fact that part of the run-off needs to be maintained for the environment. Therefore it is better to subtract the environmental flow requirement from total run-off (Smakhtin et al, 2004; Poff et al, 2010). Third, comparing water use to actual run-off from a catchment becomes problematic when run-off has been substantially lowered due to the water use within the catchment. It makes more sense to compare water use to *natural* run-off from the catchment, in other words, the run-off that would occur without consumptive water use within the catchment. Finally, it is not so accurate to consider water scarcity by comparing *annual* values of water use and availability (Savenije, 2000). In reality, water scarcity manifests itself at a monthly rather than annual scale, due to the intra-annual variations of both water use and availability. In the context of water footprint studies, the 'blue water scarcity' in a catchment is defined such that the four weaknesses are repaired.

Environmental sustainability of the grey water footprint

The effect of the total grey water footprint in a catchment depends on the run-off in the catchment available to assimilate the waste. A grey water footprint in a specific period in a specific catchment forms a hotspot when ambient water quality standards in that period in that catchment are violated, in other words, when waste assimilation capacity is fully consumed.

As a relevant local impact indicator, one can calculate the 'water pollution level' (WPL) within a catchment, which measures the degree of pollution. It is defined as the fraction of the waste assimilation capacity consumed and calculated by taking the ratio of the total of grey water footprints in a catchment (ΣWF_{grey}) to the actual run-off from that catchment (R_{act}). A water pollution level of 100 per cent means that the waste assimilation capacity has been fully consumed. When the water pollution level exceeds 100 per cent, ambient water quality standards are violated. The water pollution level is thus calculated for a catchment x and time t as follows:

$$WPL[x,t] = \frac{\sum WF_{grey}[x,t]}{R_{act}[x,t]} \quad [-] \tag{55}$$

Both the grey water footprint and the run-off vary within the year, so that the water pollution level will fluctuate within the year as well. In most cases, calculation per month is probably good enough to represent the variation in time; if necessary, it is possible of course to take a smaller time step. One can calculate the water pollution level for larger as well as for smaller catchment areas. The disadvantage of assessing the water pollution level for relatively large catchment areas at once, is that one calculates an average for the catchment as a whole, which means that the outcome will not be able to show differences in pollution level within the catchment. This is nicely illustrated in the second calculation example in Appendix IV.

In summary, the indicators of green and blue water scarcity and water pollution level have been defined such that when values exceed 100 per cent this reflects unsustainable conditions and indicates an environmental hotspot. An environmental hotspot is defined as a period of time in a catchment in which environmental green or blue water needs or water quality standards are violated.

4.2.3 Social sustainability criteria for identifying social hotspots

The total water footprint in a catchment is socially unsustainable and thus creates a social hotspot when basic human needs are not met by all people in the catchment or when basic rules of fairness are not met, provided that the water footprint in the catchment can be seen as partially related to that. Water-related basic human needs include a minimum amount of safe and clean freshwater supply for drinking, washing and cooking (UN, 2010b) and a minimum allocation of water to food production to secure a sufficient level of food supply to all. The latter right, the 'right to water for food' has not been formally established, but food itself as a human right has already been established in the Universal Declaration of Human Rights (UN, 1948). Another basic human need is employment, which may be at stake when, for example, downstream fishermen are affected by pollution from upstream. Basic rules of fairness include the water user pays principle and the polluter pays principle. It is not fair and therefore not sustainable if some upstream people have a blue or grey water footprint that leads to problems for downstream people when those downstream people are not properly compensated by the upstream water users and polluters. Another rule of fairness is fair use of public goods. Since freshwater is basically a public good, it may be considered unfair for example when some users consume more than a fair share from an aquifer or freshwater lake. An example is when commercial farmers dig deep wells to irrigate their crops, making it more difficult for surrounding smallholder farms to get access to water.

Basic human needs and rules of fairness are criteria that are difficult to quantify in the form of sharp boundaries. Whether water-related basic human needs or rules of fairness in a certain catchment are violated will depend on expert judgement. The existence of conflicts over water can be a practical indication (Gleick, 2010; Oregon State University, 2010). In practice, social conflicts over water will often arise at the same time as when environmental conflicts occur. Therefore, the identification of environmental hotspots will also generate a list of potential social hotspots.

4.2.4 Economic sustainability criteria for identifying economic hotspots

The total water footprint in a catchment is economically unsustainable and thus creates an economic hotspot when water is not allocated and used in an economically efficient way. The benefits of a (green, blue or grey) water footprint that results from using water for a certain purpose should outweigh the full cost associated with this water footprint, including externalities, opportunity costs and a water scarcity rent. Water in a catchment needs to be allocated in an economically efficient way to different users (allocation efficiency) and each user should use its allocated water efficiently as well (productive efficiency). When the price of water for the user is below its real economic cost, this often results in inefficient use, so the degree of charging full economic costs to the water user can be an indicator.

4.2.5 Assessing primary and secondary impacts in the hotspots identified

With the hotspots we know in which catchments and in which periods of the year water scarcity and water pollution conflict with environmental, social and economic criteria of sustainability. We also know the severity in the hotspots, because the larger the green or blue water scarcity or the larger the water pollution level, the greater the problem. After having localized the hotspots and having established the severity per hotspot, one can assess the primary and secondary impacts per hotspot in more detail if this is part of the scope set for the assessment.

Showing the primary impacts in the hotspots can be done at different levels of detail. One can apply a simple water balance model or an advanced physical-based hydrological model or anything in between to estimate the effect of the green and blue water footprint on the catchment hydrology. Water quality models are also available in all forms, from simple models requiring few inputs to advanced models with high data requirement. The most important primary impact variables are: run-off and associated water levels, and some water quality parameters relevant to the case studied. In order to be meaningful, all variables need to be compared with baseline hydrological and water quality conditions. As a baseline one can best take the natural conditions within the catchment, as in undeveloped state. In this way one can visualize the full human impact.

When we speak about assessing the secondary impacts of green, blue and grey water footprints, we enter a field where much literature is available already, although structuring an environmental, social or economic impact assessment is still a major challenge. For a broad assessment, one explicitly distinguishes between environmental, social and economic impacts. The first question is generally what impact variables are to be taken into account. Textbooks on impact assessment generally provide long lists of impact variables to be included. Environmental variables generally include parameters such as abundance of certain species, biodiversity, loss of habitat. Social variables often include things like human health, employment, distribution of welfare and food security. Economic variables will include income in different sectors of the economy (in the case of reduced water flows or deteriorated water quality, specific sectors such as fishing, tourism, hydropower generation and navigation may suffer). All the time, it is a challenge to make the secondary impact variables measurable. When it is clear what list of secondary impact variables is going to be used, the second question is how to estimate how the primary impacts (changed water flows and water quality) can be translated into reliable estimates of secondary impacts. One can use models, expert judgement and participatory approaches to make this step. We suffice here with referring the reader to the vast and diverse amount of literature that exists in the field of impact assessment.

4.3 Sustainability of the water footprint of a process

Whether the water footprint of a specific process is sustainable or not depends on two criteria:

1. *Geographic context:* the water footprint of a process is unsustainable when the process is situated in a hotspot, in other words, in a certain catchment area in a certain period of the year in which the total water footprint is

unsustainable from either environmental, social or economic point of view.

2. *Characteristics of the process itself:* the water footprint of a process is unsustainable in itself – independent of the geographic context – when the water footprint of the process can be reduced or avoided altogether (at acceptable societal cost).

The two criteria need to be evaluated separately for the green, blue as well as for the grey water footprint. The first criterion simply implies that when the water footprint of a process contributes to a hotspot, where the *overall* water footprint is unsustainable, the water footprint of this particular process is unsustainable as well. As long as the total water footprint in a catchment in a specific period is unsustainable, every specific contribution is to be regarded as unsustainable, even though the contribution may be relatively small. This notion is based on the recognition that there is shared risk and responsibility. If the total water footprint in a catchment is unsustainable – for example, because the blue water footprint exceeds the blue water availability – one cannot single out one component in the total that forms the problem, since it is the total that creates the problem. When the water footprint of a process contributes to a hotspot, it is unsustainable because it is part of an unsustainable situation. How to define hotspots from environmental, social or economic point of view has been extensively discussed in the previous section. Therefore in this section we need to further explain only the second criterion.

The green, blue or grey water footprint of a process is unsustainable in itself when the footprint can be avoided or reduced because better technology is available at acceptable societal cost. This can be the case in water-scarce catchments, but in water-abundant catchments as well. Many processes can be improved or replaced by other processes that have no or much lower water footprint at a reasonable societal cost or even with a resulting societal benefit. We are inclined to think that lowering water footprints will cost money (wastewater treatment, more efficient irrigation techniques, measures to bring about more efficient use of rainwater), but this is generally a finding that results from a micro-perspective, from the view of the one that has to initially invest in the necessary measures. From a macro-point of view, when internalizing economic and environmental externalities posed by overexploitation and pollution of water resources, water footprint reduction will generally result in a societal benefit or, at most, a reasonable societal cost.

Most forms of water pollution are unnecessary and avoidable. Therefore, nearly all processes that result in a grey water footprint are unsustainable. Many processes with a blue water footprint are also unsustainable. In industries, only when freshwater needs to be incorporated into a product, the associated blue water footprint cannot be avoided, but blue water footprints that refer to water evaporation in industrial processes can generally be avoided by recapturing the water. An unsustainable process is, for instance, cooling with water without capturing the evaporated water for reuse. In agriculture, blue water footprints are unsustainable when inefficient irrigation techniques are used that result in unnecessary additional evaporation.

Processes that are thus described as 'unsustainable' are not necessarily causing immediate problems of water scarcity or pollution within the catchment or river basin in which they occur (for example, when there are few other water users, so that environmental flow requirements are still met and waste assimilation capacity not yet fully consumed), but nevertheless they are not sustainable, because they unnecessarily consume water and usurp waste assimilation capacity. When green and blue water footprints in water-abundant areas are unnecessarily large, this generally reflects low water productivity, in other words, low product output per unit of water consumption. This is unsustainable given the fact that water productivities need to be increased in water-abundant areas in order to reduce the necessity to produce water-intensive products in water-scarce areas.

Unfortunately, no clear criteria exist yet to determine whether a process is unsustainable in itself, so that for the time being one will have to depend on judgement by experts based on available techniques. Global benchmarks need to be developed, so that one can compare the water footprint of a specific process to a global benchmark for that process. The benchmark should indicate the 'reasonable' maximum water footprint per unit of output product derived from a process, expressed separately for the green, blue and grey water footprint.

4.4 Sustainability of the water footprint of a product

4.4.1 Identifying the unsustainable components in the water footprint of a product

The water footprint of a product is the sum of the water footprints of the process steps that were necessary to produce the product (see Sections 3.2 and 3.4). The sustainability of the water footprint of the product therefore depends on the sustainability of the water footprints of the various process steps. Each process step takes place in one or more specific catchments, often in a specific time of the year. The overall water footprint of a product thus consists of a lot of separate components, each one of which refers to one specific process and occurs in a specific part of the year in a specific catchment. Each separate component in the water footprint of a product can be assessed in terms of sustainability based on two criteria:

1. Is the water footprint component located in a catchment area and period of the year that was identified as a hotspot?
2. Is the water footprint of the process itself unsustainable: in other words, can the water footprint be avoided altogether or reduced at reasonable societal cost?

This procedure needs to be carried out separately for the green, blue and grey component of the product water footprint. The procedure is illustrated in Table 4.1, which shows a hypothetical product. Its production system consists of six process steps. Some of the processes are located in more than one catchment. The two above-mentioned criteria are applied separately for each water footprint component. Some components have a negative score on one criterion, some components a negative score on the other criterion, some components have two negative scores and some components two positive scores. The two criteria of geographic sustainability and sustainability of the process itself complement each other. That means that each separate component of the water footprint of a product can be unsustainable either because it contributes to a geographically unsustainable situation (a hotspot) or because it refers to a process that is unsustainable in itself.

The final conclusion after having assessed the sustainability of a product water footprint can be stated in terms like 'x per cent of the water footprint of the product is unsustainable'. One can show which are the unsustainable components in the overall water footprint and one can explain why these components are unsustainable: either because they concern a water footprint that can be avoided altogether or reduced at reasonable societal cost or they contribute to a hotspot (or both). The unsustainable components in the water footprint of a product deserve action in order to improve the situation. Based on the share that a certain water footprint component has in the total water footprint of the product, one can set priorities with respect to where to start. One can even decide to disregard altogether components that are unsustainable but contribute to the overall product water footprint below a certain threshold (for example, 1 per cent). Prioritizing can also be done on the basis of the relative severity of the various hotspots to which the different unsustainable water footprint components contribute or on the basis of which improvements can most rapidly and easily be achieved.

As for individual processes, benchmarks need to be developed for products as well. In this way, one would be able to compare the water footprint of a specific product to a global benchmark for that product, which refers to a 'reasonable maximum water footprint per unit of product'. The latter can be considered as the sum of the reasonable maximum water footprints that have been established earlier for each process step in the production system of the product.

Table 4.1 *Example of how to assess the extent to which the water footprint of a product is sustainable, based on two criteria: geographic sustainability of the water footprints in the catchments in which the process steps are located and sustainability of the underlying process steps themselves. Priority components in the water footprint of a product can be identified based on which components are unsustainable and the share of a component in the total water footprint of the product. The table needs to be filled separately for the green, blue and grey water footprint of the product*

Data derived from the product water footprint account			***Check the sustainability of the total water footprint in the catchment in which the process is located***	***Check the sustainability of the water footprint of the process itself***	***Conclusion***		***Check relevance from product perspective***	***Check whether response is required***
Process step[a]	***Catchment in which the process is located***[b]	***Water footprint (m^3 per unit of final product)***	***Is the catchment a hotspot?***	***Can the water footprint be reduced or avoided altogether?***	***Is this a sustainable component in the product water footprint?***	***Fraction of the product water footprint that is not sustainable***	***Share above threshold of one per cent***[c]	***Is this a priority component?***
1	A	45	no	no	yes		yes	no
	B	35	**yes**	**yes**	**no**	35%	yes	**yes**
2	A	10	no	no	yes		yes	no
3	C	6	no	no	yes		yes	no
	D	2	**yes**	no	**no**	2%	yes	**yes**
	E	1.1	no	**yes**	**no**	1.1%	yes	**yes**
4	F	0.5	**yes**	no	**no**	0.5%	no	no
5	A	0.3	no	no	yes		no	no
6	A	0.1	no	**yes**	**no**	0.1%	no	no
total		100				38.7%		

[a] The production system of the product consists of a number of sequential or parallel process steps (see Section 3.4.2).

[b] A process step (for example, growing a particular crop which is an ingredient to the product considered) can be located in different catchments.

[c] Choosing the threshold can be subject to debate.

4.4.2 Water footprint impact indices reflecting local environmental impacts

A detailed assessment of the sustainability of the water footprint of a product as summarized in Table 4.1 is useful for understanding where it hurts most, so that dedicated measures can be formulated. For some purposes, most notably for the purpose of carrying out a life cycle assessment (LCA), it is desired to summarize the information on the sustainability of a product water footprint into one index or a few indices. LCA studies aim to assess the overall environmental impact of products; the impacts on water resource use and water quality are just two types of environmental impact among a number of other environmental impacts of products. In LCA studies, all impacts need to be expressed in single indices, which requires aggregation of more specific information.

Freshwater resources in the world are limited and in many places water resources are already heavily overused. Measuring water resources use by looking at water consumption and pollution in terms of volumes should therefore be a key element in an LCA study. The green, blue and grey water footprints of a product are good indicators of total water resource consumption and use of waste assimilation capacity related to the product. The green, blue and grey water footprints of a product can thus directly be used as indicators in LCA. However, apart from the fact that it is relevant to look at the *volume of freshwater appropriation*, it is also interesting to look at the *local environmental impacts* related to the freshwater appropriation. This local environmental impact depends on the water scarcity and water pollution level in the catchments in which the water footprint of the product is located. As a measure of the local environmental impacts of the water footprint of a product, one can use the water footprint impact indices as described in this section. Recently, the LCA research community has very much focused on these local environmental impacts of water use, ignoring the larger issue of global water scarcity (Pfister and Hellweg, 2009; Ridoutt and Pfister, 2010). It is emphasized here that these local impacts are just part of the water issue. The issue of how much water in total is being appropriated for a product is as relevant if not even more. When two products have the same water footprint, they make a similar claim on the globe's limited water resources, even though, when made in two different places, the local environmental impact may be different.

When a certain product contributes to the total blue water footprint in a catchment, the impact of that specific water footprint depends on two factors: (i) how large the specific blue water footprint is, and (ii) what the blue water scarcity in the catchment is. For the green water footprint the same rationale can be followed. Similarly, when a certain product contributes to the total grey water footprint in a catchment, the impact of that depends on: (i) how large the specific grey water footprint is, and (ii) what the water pollution level in the catchment is.

The water footprint is a volumetric measure, showing freshwater consumption and pollution in time and space. Water footprints provide relevant information on how water resources are allocated to different purposes. The water footprint of a product shows the 'water allocated' to that product. The water allocated to one product cannot be allocated to another product. In terms of freshwater appropriation, a blue water footprint of one cubic metre is always equivalent to another blue water footprint of one cubic metre, even though the former may be in a water-scarce catchment while the latter happens to be in a water-rich catchment. For the overall picture of freshwater appropriation it does not really matter which of the two footprints is located in the water-scarce catchment, since it could have been the other way around. The two footprints have the same impact on overall freshwater appropriation. The appropriated water volume for a process or product provides key information in the allocation discussion, but does not provide information on whether it contributes to an immediate problem of water scarcity or pollution within the catchment where it occurs. In terms of local impact, a blue water footprint of $1 m^3$ is not necessarily equivalent to another blue water footprint of $1 m^3$, because one can occur in a catchment where environmental flow requirements are violated while the other one happens to be in a catchment where this is not the case.

For the purpose of visualizing the local impact, one will need to put the blue water footprint of a specific product in the context of the blue water scarcity in the catchment where the footprint occurs. Similarly, the green water footprint needs to be considered in the context of green water scarcity. The grey water footprint of a specific product in a catchment needs to be regarded in the context of the water pollution level in the catchment.

The 'green water footprint impact index' ($WFII_{green}$) is an aggregated and weighed measure of the environmental impact of a green water footprint. It is based on two inputs: (i) the green water footprint of a product specified by catchment x and by month t, and (ii) the green water scarcity by catchment and by month. The index is obtained by multiplying the two matrices and then summing the elements of the resultant matrix. The outcome can be interpreted as a green water footprint weighed according to the green water scarcity in the places and periods where the various green water footprint components occur.

$$WFII_{green} = \sum_x \sum_t \left(WF_{green}[x,t] \times WS_{green}[x,t] \right) \tag{56}$$

The 'blue water footprint impact index' ($WFII_{blue}$) is an aggregated and weighed measure of the environmental impact of a blue water footprint. It is based on: (i) the blue water footprint of a product specified by catchment x and by month

t, and (ii) the blue water scarcity by catchment and by month. The index is obtained by multiplying the two matrices and then summing the elements of the resultant matrix. The outcome can be interpreted as a blue water footprint weighed according to the blue water scarcity in the places and periods where the various blue water footprint components occur.

$$WFII_{blue} = \sum_{x}\sum_{t}\left(WF_{blue}[x,t] \times WS_{blue}[x,t]\right) \quad (57)$$

The 'grey water footprint impact index' ($WFII_{grey}$) is an aggregated and weighed measure of the environmental impact of a grey water footprint. It is based on: (i) the grey water footprint of a product specified by catchment x and by month t, and (ii) the water pollution level by catchment and by month. The index is obtained by multiplying the two matrices and then summing the elements of the resultant matrix. The outcome can be interpreted as a grey water footprint weighed according to the water pollution level in the places and periods where the various grey water footprint components occur.

$$WFII_{grey} = \sum_{x}\sum_{t}\left(WF_{grey}[x,t] \times WPL[x,t]\right) \quad (58)$$

The three water footprint impact indices refer to different sorts of water use which are not comparable. In order to have an overall water footprint impact index one could simply add the three indices. Since green water scarcity is generally lower than blue water scarcity, the green water footprints will count less than the blue water footprints.

As a general note, we would like to emphasize that the impact indices as discussed here have a limited value. The reason is that the useful information for response is contained in the underlying variables. It is relevant to know the size and colour of a water footprint, to know when and where it occurs and in which context (degree of water scarcity, water pollution level). Aggregating this information into three indices or synthesizing the three into one overall index means that all information is covered. What remains is a crude impression of the local environmental impact of a water footprint as a whole, which can be useful when one aims to roughly compare it with the local impact of another water footprint, but is not useful as a basis for formulating specific response measures. It should also be noted that the above water footprint impact indices account for environmental impacts only, not social or economic impacts. Besides, they show impacts at catchment level; for considerations of sustainable water use, the volumetric accounts provided by the water footprint indicator are more useful. The indices presented here are, however, useful for LCA studies, which require such highly aggregated impact indices.

Water footprint impact indices are useful only as crude indicators of the environmental impact at catchment level; the aggregated indices no longer contain spatial or temporal information. As a basis for formulating appropriate response measures, it is more useful to identify 'hotspots' as explained earlier than to calculate aggregated water footprint impact indices. It should also be noted that the impact indices discussed here aim to measure environmental impacts at the catchment level; for assessing sustainable water allocation, indices that reflect local impacts are not helpful. For this purpose, one can better use the volumetric water footprint accounts, because allocation is about apportioning scarce resources, not about local impacts.

4.5 Sustainability of the water footprint of a business

The water footprint of a business is the same as the sum of the water footprints of the final products that the business produces (see Sections 3.2 and 3.10). Therefore one should first assess the sustainability of the water footprints of the products produced by the producer and then assess the sustainability of the water footprint of the business. This is a minor step, because the findings with respect to the sustainability of the water footprints of the separate business products can be immediately translated to a conclusion about the sustainability of the water footprint of the business. Suppose that a business produces two final products and that three quarters of the business water footprint relates to the one product and one quarter to the other. It was found that one third of the water footprint of the first product is not sustainable and that the water footprint of the second product is unsustainable as a whole. In this case, it follows that ($3/4 \times 1/3 + 1/4 \times 1 =$) 50 per cent of the water footprint of the business as a whole is unsustainable. The water footprint sustainability assessments for the two products can be used to see the processes that are responsible for the unsustainable components in the business water footprint and to identify in which catchments these processes are located.

4.6 Sustainability of the water footprint of a consumer

The water footprint of a consumer is equal to the sum of the water footprints of the products used by the consumer. Therefore, the sustainability of the water footprint of a consumer depends on the sustainability of the water footprints of the products used. One can simply apply the method as described in Section 4.4 for all products used. In this way, one finds the sustainability for each component of the water footprint of a consumer. For assessing the sustainability of the water footprint of a consumer, looking at the sustainability of all separate water footprint components is not sufficient, however. One should also look at the water footprint as a whole. Therefore, a second criterion comes into play. The sustainability of the water footprint of a consumer also depends on whether the consumer water footprint is smaller or larger than an individual's fair share given the limitations to the total water footprint of humanity.

The water footprint of many consumers will be dominated by a few components only. For meat eaters, this will generally be the water footprint of the meat component in their consumption (Hoekstra, 2010b). For consumers with a relatively large water footprint, one may like to identify the products of most concern. Products of concern are 'luxury' products that are associated with a relatively large water footprint. Large-scale allocation of water to 'luxury' goods can be at the cost of water supply to the environment or at the cost of water supply for the production of basic food commodities. Example luxury goods with relative large water footprints other than meat include agriculture-based cosmetic products and first-generation biodiesel or bio-ethanol. These products are not immediately unsustainable when produced at a limited scale, provided that production takes place in non-hotspot catchments, but they become unsustainable as soon as at a global scale water resources allocation to these products turns out to be at the cost of allocating the world's scarce water resources to purposes that sustain basic needs. This can happen, for instance, when a cereal such as maize is increasingly being used for producing bio-ethanol for the benefit of some, while at the same time others experience reduced food security because of increased maize prices. Products of concern are associated with large water footprints that form the top-layer of humanity's water footprint. Reduction or a hold to the augmentation of this top-layer water footprint is necessary in order to preserve sufficient water to be allocated to the environment and basic human needs.

The sustainability of the water footprint of a group of consumers – for example, the consumers within a nation – depends on the sustainability of the water footprints of the individual consumers. One can hereby look at whether individual consumers have a smaller or larger water footprint than their fair share, but one can also look at whether national consumption as a whole puts a claim on the globe's limited freshwater resources that remains below or goes beyond the nation's fair share.

Chapter 5

Library of Water Footprint Response Options

5.1 Shared responsibility

One can argue that consumers are responsible for what they consume, so they are also responsible for the indirect resource use related to their consumption pattern. In this sense, consumers have responsibility for their water footprint and should undertake action to ensure that their water footprint is sustainable. If they would do so, producers would be forced to deliver sustainable products. One can also turn the argument around and argue that producers are responsible for delivering sustainable products. This would imply that producers should take action to make product water footprints sustainable. And investors, of course, should include considerations of sustainable water use into their investment decisions. Finally, water is a public good, so governments cannot withdraw from their responsibility to put proper regulations and incentives in place to ensure sustainable production and consumption. It will be maintained here that consumers, producers, investors and governments all have a shared responsibility. This chapter will review options available to consumers, producers, investors and governments to reduce water footprints and particularly mitigate impacts.

It is not the purpose here to be prescriptive. The manual does not say what to do, but is restricted to an inventory of options. Since this is a first version of such an inventory, it does by no means claim to be exhaustive. Nevertheless, it may be a helpful guide in understanding along which lines alternative response strategies can be formulated. A response strategy can be a combination of one or more of the options identified here.

5.2 Reducing the water footprint of humanity: What is possible?

Technically, both blue and grey water footprints in industries and households can be reduced to zero by full water recycling. In a closed cycle there will be neither evaporation losses nor polluted effluents. In factories or cooling systems, evaporated water can be captured and recycled or returned to the water body where it was taken from. There are a few exceptions, where the blue water footprint of a process cannot completely be reduced to zero, most notably when water is used to be incorporated into a product; this part of the blue water footprint cannot be avoided, but we are talking about minor fractions in the blue water footprint of humanity. Another exception can be when water is applied in the open air by necessity, so that some evaporation cannot be avoided. The only sort of grey water footprint that cannot always be reduced to zero is the grey water footprint related to thermal pollution, but even here heat can be partly recaptured from heated effluents from cooling systems and used for other purposes before the effluent is disposed into the environment.

In agriculture, the grey water footprint can be reduced to zero by preventing the application of chemicals to the field. It can be lowered substantially by applying less chemicals and employing better techniques and timing of application (so that less chemicals arrive in the water system by run-off from the field or by leaching). Green and blue water footprints (m^3/ton) in agriculture can generally be reduced substantially by increasing green and blue water productivity (ton/m^3). Agriculture is often focussed on maximizing land productivity (ton/ha), which makes sense when land is scarce and freshwater is abundant, but when water is scarcer than land, maximizing water productivity is more important. In the case of blue water, this implies applying less irrigation water in a smarter way, in order to give a higher yield per cubic metre of water evaporated.

Table 5.1 summarizes the possible water footprint reduction targets per water footprint component per sector. The operational blue and grey water footprints in the industrial sector can be more or less nullified. In the agricultural sector, more research is needed to formulate reasonable quantitative water footprint reduction targets. In theory, the grey water footprint can be brought back to zero through organic farming. In practice, it will be quite a challenge and require substantial time before all conventional farming can be replaced by organic farming. Further, it is roughly estimated that, over a time span of a few decades, the total blue water footprint in the world can be reduced by half, to be achieved partly by increasing blue water productivities in irrigated agriculture (through the application of water-saving irrigation techniques and by 'deficit' instead of 'full' irrigation) and partly by increasing the fraction of production that is based on green instead of blue water.

'Reduction' of water footprints can be achieved in two different ways. In a certain production chain, one can replace one technique by another technique that results in a lower or even zero water footprint or one can avoid the use of a specific ingredient or final product altogether. Examples of improved production technology are: replace sprinkler irrigation by drip irrigation;

Table 5.1 *Possible water footprint reduction targets per sector and water footprint component*

	Agriculture	*Industry*
Green water footprint	Decrease green water footprint (m^3/ton) by increasing green water productivity (ton/m^3) in both rain-fed and irrigated agriculture. Increase total production from rain-fed agriculture.	Not relevant.
Blue water footprint	Decrease blue water footprint (m^3/ton) by increasing blue water productivity (ton/m^3) in irrigated agriculture. Decrease ratio blue/green water footprint. Decrease global blue water footprint (for example, by 50 per cent).	Zero blue water footprint: no losses through evaporation – full recycling – only blue water footprint related to the incorporation of water into a product cannot be avoided.
Grey water footprint	Reduced use of artificial fertilizers and pesticides; more effective application. Grey water footprint can go to zero through organic farming.	Zero grey water footprint: no pollution – full recycling, recapturing heat from heated effluents and treatment of remaining return flows.

replace conventional by organic farming; or replace open-system water cooling by closed-system water cooling. Examples of avoidance are: replace a heavy-meat meal by a vegetarian or light-meat meal (using other protein sources that are much less water-intensive); avoid the application of toxic chemicals that, through the sewer, end up in the surface or groundwater; or avoid the use of water-intensive biofuels (but use instead, for example, electricity sourced from solar or wind energy). 'Reduce by avoid' is generally more fundamental than 'reduce by improved production', because 'avoid' often requires a reconsideration of the production and consumption pattern itself, while 'improved production' allows us to do the same as before but just more 'eco-efficiently'. When exploring options to reduce water footprints it is always essential to explore both avenues.

It is often thought that water footprint reduction is only relevant in locations where problems of water scarcity and pollution exist. The idea is that it is unnecessary to reduce the blue water footprint in an area where blue water is abundantly available and needless to reduce the grey water footprint when sufficient water is available to dilute pollutants such that concentrations still remain below maximum allowable concentrations. Similarly, it is often thought that it is unnecessary to reduce the green water footprint in agriculture because the rain comes anyway and will otherwise remain unproductive. The rationale behind this thinking is: when the water footprint in a certain catchment in a certain period does not lead to significant water depletion or pollution, the water footprint must be sustainable. This sort of thinking, however, is based on the misconception that the sustainability of water use depends on the local geographic context only. As explained in Section 4.3, the water footprint of a specific process is unsustainable and therefore needs to be reduced when (i) the water footprint of the process contributes to a hotspot, or (ii) the water footprint can be reduced or avoided altogether, independent of the geographic context. The second criterion implies that water footprints need to be reduced also in water-abundant areas whenever reasonably possible, not to solve local water problems in these areas, but to contribute to a more sustainable, fair and efficient water use globally. Reducing water footprints (m^3/ton) in water-abundant areas by increasing water productivity (ton/m^3) is key in reducing the pressure on water resources in water-poor areas, since increased production of water-intensive goods in areas where water is sufficiently available is necessary when the limits to production are already exceeded in water-poor areas.

The concern should be the total water footprint of humanity. The fact that this footprint is too large becomes manifest in the hotspot areas, where local problems of water depletion and pollution are visible during certain parts of the year. Reducing the water footprint in those hotspots is an obvious necessity. But it is only half of the story. Counter-intuitively, the solution for the problems in water-scarce areas greatly lies in the water-abundant areas. In water-abundant areas one can often encounter low water productivities in rain-fed agriculture (large green water footprints). Increasing the water productivity (in other words, reducing the green water footprint) in rain-fed agriculture in water-abundant areas increases global production, thus reducing the need for water-intensive products from water-scarce areas, and therefore helping to relieve the pressure on the blue water resources in those water-scarce areas. So, from a global perspective, water footprints per ton of product need to be reduced everywhere when possible, also in water-abundant areas.

From a global point of view, reducing the water footprint by 1m^3 in one catchment is equivalent to reducing the water footprint by the same amount in another catchment, even if one catchment shows a much higher water scarcity or water pollution level than the other one. The reason is that – given the limited freshwater resources available worldwide – any reduction contributes to reducing the total resource demand. When, in water-rich catchments, one can produce more water-intensive goods with the same volume of water, the production of water-intensive goods in water-poor areas can be lessened, so that the total water footprint in those water-poor areas can be reduced. This is an indirect but important way of solving the pressure on the water resources in water-poor areas. From a local and more immediate point of view, it matters of course whether one reduces the water footprint by 1m^3 in a water-scarce or water-rich catchment. Reducing the water footprint in a water-scarce area immediately

Table 5.2 *Priorities in water footprint reduction*

	Non-hotspots	***Hotspots****
Little reduction potential	0	+
Large reduction potential**	+	++

* A hotspot is a specific period of the year (for example, the dry period) in a specific (sub)catchment in which the water footprint is unsustainable, for instance because it compromises environmental water needs or water quality standards or because the water allocation and use in the catchment is considered unfair and/or economically inefficient.
** There is a large reduction potential when the water footprint can be avoided altogether or substantially reduced at reasonable societal cost.

contributes to lessening the pressure, provided that the reduced water footprint per unit of production is not nullified by increased production at the same time. So, although all water footprint reductions contribute to solving the global problem of limited freshwater resources, one can argue that priority is to be given to reducing water footprints that are located in hotspots, because acting in hotspots has both a global and local rationale, while acting in non-hotspots has a global rationale only. Table 5.2 summarizes in a very schematic way how priorities in water footprint reduction can be set.

When talking about water footprint reduction to zero, one could use the term 'water neutral', a term chosen as an analogue to 'carbon neutral', a term that applies to activities with a carbon footprint of zero. The term water neutral can be interpreted in different ways, which makes it a bit of a problematic concept (Box 5.1). It is clearest when it simply refers to 'zero water footprint', which is possible in the case of the operations of industries, where a water footprint of zero is technically possible. The concept of water neutrality becomes less clear when it includes a form of 'offsetting'. The idea of 'water footprint offsetting' is obviously an idea developed in analogue to the concept of 'carbon offsetting'. Whereas 'carbon offsetting' already gives rise to arguments about what it precisely entails, the concept of 'water footprint offsetting' is probably even more contentious (Box 5.2). We recommend giving priority to setting quantitative targets with respect to the reduction of water footprints and associated impacts, rather than offsetting.

5.3 Consumers

The water footprint of a consumer is sustainable when (i) the total remains below the consumer's fair share in the world, (ii) no component of the total water footprint is located in a hotspot, and (iii) no component of the total water footprint can be reduced or avoided altogether at reasonable societal cost.

Box 5.1 *Water neutrality*

'Water neutral' is an idea similar to 'carbon neutral'. However, just as in the case of carbon neutrality, the term water neutrality is arguable. The problem already lies in its definition. When used for the first time, at the 2002 Johannesburg World Summit for Sustainable Development, the idea was to quantify the water consumed during the conference by delegates and translate this into money. Delegates, corporations and civil society groups were encouraged to make the summit 'water neutral' by purchasing water-neutral certificates to offset their water consumption during the ten-day summit, with the offset investment being earmarked for the installation of pumps to water needy communities in South Africa and for water conservation initiatives (Water Neutral, 2002). In 2007, the Coca-Cola Company made a pledge to become water neutral in its operations, where 'water neutral' refers to (i) reduce water use in its own operations, (ii) return the water used in its own operations in clean form back to the environment and (iii) compensate for the water that is contained in the finished beverages through conservation and recycling programmes. Also in 2007, the UK Housing Ministry released official details on the expansive redevelopment project 'Thames Gateway' that would be water-neutral, in this context meaning that the region will not require the use of additional water despite the plethora of new homes being built and people moving in. This could be done by offsetting additional water demands in the new buildings by reducing water use in existing buildings (Environment Agency, 2007). In all three cases water use is measured in terms of water abstractions (not water consumption) and all three cases include some form of 'offsetting'. Besides, all three cases consider direct water use only, not indirect forms of water use. However, the three applications of the water neutral concept put different weights to 'reducing water use' versus offsetting. The Johannesburg and Thames Gateway cases are essentially about offsetting, while the Coca-Cola case uses 'offsetting' only for that part of the water consumption that really cannot be reduced (the water that goes into the beverages). The Johannesburg and Thames Gateway cases differ in that the latter seeks offsetting within the same clearly defined area, while the former is not so sharp in that respect. Hoekstra (2008a) proposed to relate the concept of water neutral to the water footprint concept and to define it as: reduce the water footprint of an activity or product as much as reasonably possible and offset the negative externalities of the remaining water footprint. By relating the water neutral concept to the water footprint it now addresses indirect water use as well. In some particular cases, when interference with the water cycle can be completely avoided – for example, by full water recycling and zero waste – 'water neutral' according to this definition means that the water footprint is nullified; in many other cases, such as in the case of crop growth, water consumption cannot be nullified. Therefore, with this definition, 'water neutral' does not always mean that water consumption is brought down to zero, but that the water footprint is reduced as much as possible and that the possible impacts of the remaining footprint are fully compensated. Compensation can be done by contributing to (investing in) a more sustainable and equitable use of water in the hydrological units in which the impacts of the remaining water footprint are located. In the latest definition of the water neutral concept, a number of important questions remain. These are, for example: How much reduction of a water footprint can reasonably be expected? What is an appropriate water-offset price? What type of efforts count as an offset? As long as these sorts of questions have not been answered yet, the risk of the water neutrality concept is that its content depends on the user. As a result, some may use it to refer to real good measures taken in both operations and supply chain while others may use it only as an attractive term in an advertisement. The risk with the water-neutral concept is also that the focus will shift from water footprint reduction to offsetting. A water footprint can be measured empirically, so can its reduction. Defining offsetting and measuring its effectiveness is much more difficult, enlarging the risk of misuse. Besides, compensating measures should be considered a last resort option, to be looked at after having reduced the company's water footprint first.

Box 5.2 *Water footprint offsetting*

The concept of water footprint offsetting is still ill-defined. In general terms it means taking measures to compensate for the negative impacts of the water footprint that remains after reduction measures have been implemented. But the two weak points in the definition are that (i) it does not specify which sort of compensation measures and which level of compensation are good enough to offset a certain water footprint impact and (ii) it does not specify which impacts should be compensated precisely and how to measure those impacts. In Chapter 4, we have seen that the term 'impact' can be interpreted very broadly. The fact that the offset concept is ill-defined means that it can easily be misused. Without a clear definition, measures taken under the banner of 'offset' can potentially be a form of 'greenwashing' rather than a real effort aimed at full compensation. For this reason, we strongly recommend focusing response on reducing water footprints and looking at offsetting as a real last step only. Another reason is that water footprints and their associated impacts are always local. In this respect, the water footprint is markedly different from the carbon footprint. The idea of a global offset market as has developed over the past few years for carbon footprint offsets does not make sense for water. An offset of a water footprint should always occur in the catchment where the water footprint is located. This drives the attention to a company's own water footprint again and does not allow to think in terms of general compensation schemes where one can simply 'buy' an offset.

Consumers can reduce their direct water footprint (home water use) by installing water-saving toilets, applying a water-saving showerhead, turning off the tap during teeth-brushing, using less water in the garden and by not disposing of medicines, paints or other pollutants through the sink.

The indirect water footprint of a consumer is generally much larger than the direct one. A consumer has basically two options to reduce his or her indirect water footprint. One option is to change the consumption pattern by substituting a specific consumer product that has a large water footprint by another type of product that has a smaller water footprint. Examples include: eating less meat or becoming vegetarian, drinking plain water instead of coffee, or wearing less cotton and more artificial fibre clothes. This approach has limitations, because many people do not easily shift from eating meat to being vegetarian and people like their coffee and cotton. A second option is to select the cotton, beef or coffee that has a relatively low water footprint or that has its footprint in an area that does not have high water scarcity. This requires, however, that consumers have the proper information to make that choice. Since this information is generally not easily available, an important thing consumers can do now is ask product transparency from businesses and regulation from governments. When information is available on the impacts of a certain article on the water system, consumers can make conscious choices about what they buy.

5.4 Companies

A corporate water footprint strategy can include a variety of targets and activities (Table 5.3). Businesses can reduce their operational water footprint by reducing water consumption in their own operations and bringing water pollution to zero. Keywords are: avoid, reduce, recycle and treat before disposal. By avoiding any evaporation, the blue water footprint can be reduced to zero. By reducing the production of wastewater as much as possible and by treating the wastewater still produced, the grey water footprint can be reduced to zero as well. Treatment can be done within the company's own facilities or by a public wastewater treatment facility; it is the quality of the water finally discharged into the ambient water system that determines the grey water footprint.

For most businesses, the supply chain water footprint is much larger than the operational footprint. It is therefore crucial that businesses address that as well. Achieving improvements in the supply chain may be more difficult – because not under direct control – but they may be more effective. Businesses can reduce their supply chain water footprint by making supply agreements including certain standards with their suppliers or by simply changing to another supplier. In many cases this may be quite a task, because the whole business model may need to be transformed in order to incorporate or better control supply chains and to make supply chains fully transparent to consumers.

A company can also aim to reduce the consumer water footprint that is inherent to the use of their product. When consumers use a soap, shampoo, cleaning chemical or paint, it is likely that they will flush it through the drain. When the water is not treated or when the chemical is such that it is not or only partly removed, this will give a grey water footprint that could have been avoided when the company had used substances that are less toxic, less harmful and more easily degradable.

Among the various alternative or supplementary tools that can help improving transparency are: conform to shared definitions and methods (as contained in this manual), water footprint reporting and disclosure of relevant data. Clarity about activities undertaken to reduce the corporate water footprint can be enhanced by setting quantitative water-footprint reduction targets in time. A potential tool within large companies or in specific sectors is benchmarking: what can be achieved in (the supply chain of) one factory should also be possible in (the supply chain of) another factory.

5.5 Farmers

Farming is a sort of business for which the same things apply as discussed in the previous section. For livestock farmers, a major concern should be the water footprint of the feed they buy or produce themselves. For crop farmers, a number of specific water footprint reduction options are available as listed in Table 5.4. The advantage of reducing green water footprints per unit of crop in rain-fed agriculture is that total production from rain-fed lands increases. As a result of increased rain-fed production, less needs to be produced elsewhere, lessening the claims on land and (green or blue) water resources elsewhere. Reducing the green water footprint per ton of crop in one place can thus result in a reduction of the blue water footprint in crop production as a whole. In irrigated agriculture, changing irrigation technique and application philosophy can greatly reduce the blue water footprint. Using drip irrigation instead of sprinkler or furrow irrigation can reduce evaporation substantially. Further, the conventional farmer strategy to optimize yields (ton/ha) often leads to unnecessary use of irrigation water. Instead of applying full irrigation, it may be wiser to choose for so-called 'deficit irrigation', an irrigation philosophy that aims at obtaining maximum crop water productivity (ton/m^3) rather than maximum yields (ton/ha). In deficit irrigation, water is applied during the drought-sensitive growth stages of a crop; outside these periods, irrigation is limited or even unnecessary if rainfall provides a minimum supply of water. Another alternative is so-called

Table 5.3 *Corporate water footprint response options*

Water-footprint reduction targets – operations

- Benchmarking products or sites. Define best practice and formulate targets to achieve best practice throughout the business. Can be done in own company or within a sector as a whole.
- Reduction of blue water footprint in general. Reduction of consumptive water use in operations by recycling, adopt water-saving appliances, replace water-intensive by water-extensive processes.
- Reduction of blue water footprint in hotspots. Focus above measures in water-scarce areas or in areas where environmental flow requirement in a river are violated or where groundwater or lake levels are dropping.
- Reduction of grey water footprint in general. Reduce wastewater volume; recycle chemicals. Wastewater treatment before disposal. Recapture heat from wasteflows.
- Reduction of grey water footprint in hotspots. Focus above measures in areas where ambient water quality standards are violated.

Water-footprint reduction targets – supply chain

- Agree on reduction targets with suppliers.
- Shift to other supplier.
- Get more or full control over the supply chain. Change business model in order to incorporate or get better control over the supply chain.

Water-footprint reduction targets – end use

- Reduce inherent water requirements in use phase. Reduce expected water use when product is used (for example, dual flush toilets, dry sanitation equipment, water-saving showerheads, water-saving washing machines, water-saving irrigation equipment).
- Reduce risk of pollution in use phase. Avoid or minimize the use of substances in products that may be harmful when reaching the water (for example, in soaps, shampoos).

Water-footprint offsetting measures

- Environmental compensation. Invest in improved catchment management and sustainable water use in the catchment where the company's (residual) water footprint is located.
- Social compensation. Invest in equitable water use in the catchment where the company's (residual) water footprint is located, for example, by poverty alleviation and improved access to clean water supply and sanitation.
- Economic compensation. Compensate downstream users that are affected by intensive upstream water use in the catchment where the company's (residual) water footprint is located.

Product and business transparency

- Conform to shared definitions and methods. Promote and adopt the global standard for water footprint accounting and assessment as laid down in this manual.
- Promote water accounting over the full supply chain. Cooperate with others along the supply chain to be able to produce full accounts for final products.
- Corporate water footprint reporting. Report water-related efforts, targets and progress made in annual sustainability report, also covering the supply chain.
- Product water footprint disclosure. Disclosure of relevant data through reporting or internet.
- Product water labelling. Same as above, but now putting the information on a label, either separate or included in a broader label.
- Business water certification. Promote and help setting up a water certification scheme and conform to it.

Engagement

- Consumer communication; engagement with civil society organizations.
- Proactively work with governments on developing relevant regulation and legislation.

Table 5.4 *Options for crop farmers to reduce their water footprint*

Reduce green water footprint in crop growth • Increase land productivity (yield, ton/ha) in rain-fed agriculture by improving agricultural practice; since the rain on the field remains the same, water productivity (ton/m^3) will increase and the green water footprint (m^3/ton) will reduce. • Mulching of the soil, thus reducing evaporation from the soil surface.
Reduce blue water footprint in crop growth • Shift to an irrigation technique with lower evaporation loss. • Choose another crop or crop variety that better fits the regional climate, so needs less irrigation water. • Increase blue water productivity (ton/m^3) instead of maximizing land productivity (yield, ton/ha). • Improve the irrigation schedule by optimizing timing and volumes of application. • Irrigate less (deficit irrigation or supplementary irrigation) or not at all. • Reduce evaporation losses from water storage in reservoirs and from the water distribution system.
Reduce grey water footprint in crop growth • Apply less or no chemicals (artificial fertilizers, pesticides), for example, organic farming. • Apply fertilizers or compost in a form that allows easy uptake, so that leaching and run-off are reduced. • Optimize the timing and technique of adding chemicals, so that less is needed and/or less leaches or runs off.

'supplementary irrigation', which saves even more water. In this strategy, small amounts of water are added to essentially rain-fed crops during times when rainfall fails to provide sufficient moisture for normal plant growth, in order to improve and stabilize yields. The grey water footprint in the operations of the farmer can be largely reduced by adopting organic farming, which excludes or strictly limits the use of manufactured fertilizers, pesticides and other chemicals.

5.6 Investors

Not explicitly addressing the water footprint of a business and formulating appropriate response (see the previous section) may translate in various sorts of business risk (Levinson et al, 2008; Pegram et al, 2009; Morrison et al, 2009, 2010; Barton, 2010). First, there is the physical risk that companies may face freshwater shortage affecting their supply chain or own operations. Second, the corporate image of a company may be damaged in a situation where questions are raised among the public and in the media regarding whether the company properly addresses issues of sustainable and equitable water use. Problems of water depletion or pollution in the supply chain or operations of a company and the lack of mitigating strategies constitute a reputational risk for a company. Third, triggered by the wish to achieve a more sustainable and equitable use of scarce freshwater resources, governmental interference and regulation in the area of water use will undoubtedly increase. Uncertainty about future regulatory control constitutes a risk for companies that they can better anticipate than ignore. Each of the three above-mentioned risks may translate to a financial risk in terms of increased costs and/or reduced revenues. Hence investors are becoming more and more interested in the disclosure of information on the water-related risks of the business they invest in.

Risk can actually turn into opportunity for those companies that proactively respond to the challenge of global freshwater scarcity. Front-runners that create product transparency before others do, that formulate specific and measurable targets with respect to water footprint reduction – with special attention to areas where problems of water scarcity and pollution are most critical – and that can demonstrate actual improvements, can turn this into a competitive advantage.

Finally, apart from the need to address risks and the opportunity to profit from a proactive strategy, addressing the issues of freshwater scarcity and pollution should be seen as part of the corporate social responsibility. Currently, environmental concerns in companies are mostly related to energy issues. Expanding the attention towards the field of freshwater is a matter of logic in a world where freshwater scarcity is generally mentioned as the other big environmental challenge next to global warming.

5.7 Governments

Developing and implementing good water policy is only one part of wise water governance. Wise water governance also requires that governments translate the goal of sustainable water resources use to what it means for other policy fields. The governmental aim to use freshwater resources in a way that is environmentally sustainable, socially equitable and economically efficient, needs to be reflected in the government's water policy, but also in the government's environmental, agricultural, industrial, energy, trade and foreign policy. Coherence between different sector policies is crucial, because good 'water policy' in its more narrow, conventional meaning has no impact if it is undermined by, for example, an agricultural policy that leads to the aggravation of water demand in a water-scarce area. In addition, coherence is needed at different levels of governance, from the local to the national level, but international cooperation will be crucial as well. A national policy to implement proper water pricing structures in agriculture, for example, is vulnerable to failure if – in an international context – it is not agreed that other countries will develop similar policies, because of the

Table 5.5 *Options for governments to reduce water footprints and mitigate related impacts*

Water policy at national, river basin and local level

- Adopt the national water footprint accounting scheme to broaden the knowledge base for making well-informed decisions. Use information on water footprints and virtual water trade to support the formulation of both national water plans and river basin plans.
- Increase the water use efficiency at the user level, in all sectors, by promoting techniques that enlarge water productivities and thus reduce water footprints per unit of production.
- Increase the water use efficiency at the river basin level by allocating water resources to the purposes with highest societal benefit.
- Promote the allocation of available domestic water resources such that the country produces goods for which it has a comparative advantage relative to other countries.
- For national water-saving: decrease the virtual water export, increase the virtual water import and reduce the water footprint within the nation (Allan, 2003; Chapagain et al, 2006a).
- For reducing national water dependency: reduce the external water footprint.

National environmental policy

- For sustainable production: formulate reduction targets regarding the water footprint within the nation, to be specified by catchment area; focus on hotspots where impacts are largest. Translate catchment targets to operational plans involving the relevant sectors.
- For sustainable consumption: set targets with respect to the reduction of the internal and external water footprint of national consumption; focus on hotspots. Translate targets to specific product categories and economic sectors.
- Translate nature protection and biodiversity preservation goals into blue and green environmental water needs; integrate environmental water needs into river basin planning (Dyson et al, 2003; Acreman and Dunbar, 2004; Poff et al, 2010).
- Engage with consumer, environmental and other sorts of civil society organizations in helping to raise 'water awareness' among consumers, farmers and company leaders.
- Set targets with respect to the reduction of wastage in the entire food chain (from field to household losses) and formulate appropriate measures. This loss of food is equivalent to a loss in water.

National agricultural policy

- Include the goal of sustainable use of available domestic water resources in formulating national food security policy.
- Do not subsidize water-intensive agriculture in water-scarce areas.
- Promote crops that are suitable and adapted to the local climate in order to reduce irrigation demand.
- Support investments in irrigation systems and techniques that conserve water.
- Promote farmers to avoid or reduce the use of fertilizers, pesticides and insecticides or to better apply so that less chemicals reach the water system.
- Promote water footprint reduction in agriculture – see Table 5.4. This can be done in various alternative or complementary ways: regulation or legislation (for example, on timing, volumes and techniques of irrigation and on application of chemicals), water use licences, quota, full-cost water pricing, tradable water use permits, subsidies for specific irrigation techniques, compulsory water metering, awareness-raising.

Table 5.5 *Options for governments to reduce water footprints and mitigate related impacts (continued)*

National industrial / economic policy
- Promote product transparency. Implement by means of voluntary agreements by sector or by legislation.
- Translate national targets on water footprint reduction to specific reduction targets for products, producers and/or sectors. Implement through legislation and/or economic incentives (water footprint tax, and/or subsidies to specific water footprint reduction measures).

National energy policy
- Study the implications of energy scenarios for water demand, with particular attention to the water footprint of bio-energy.
- Harmonize water and energy policies so that energy policies do not increase the water footprint of the energy sector and that water policies do not increase the energy use and carbon footprint of the water sector.

National trade policy
- Ensure coherence between trade and water policies. Reduce export of low-value water-intensive products from water-scarce areas (and increase import). Use local water abundance as a factor to promote production of water-intensive goods for export.
- Reduce virtual water import dependency (in other words, reduce external water footprint) if considered necessary from a national security perspective.
- Promote product transparency of traded products, so that the water footprint of products can be traced back.

National foreign policy and international cooperation
- Promote an international agreement on worldwide water footprint reduction, for example, in the form of an international 'water footprint permit protocol' setting targeted maximum water footprints for individual countries (Hoekstra, 2006, 2010a; Hoekstra and Chapagain, 2008; Verkerk et al, 2008).
- Promote an international agreement on product transparency (Hoekstra, 2010a, 2010b).
- Promote an international water pricing protocol (Hoekstra, 2006, 2010a; Hoekstra and Chapagain, 2008; Verkerk et al, 2008).
- Cooperate with governments and other agents in developing countries to reduce water footprints; focus on hotspots in the world where water scarcity and pollution problems are most severe and where the nation contributes through its own external water footprint.

Reduce the water footprint of governmental organization and services
- See the options provided for business, Table 5.3.
- Include the water footprint of products as a criterion in government's sustainable procurement policy.

risk of unfair competition. As another example, achieving product transparency requires international cooperation as well, because many supply chains of water-intensive commodities are truly international.

Traditionally, countries formulate national water plans by looking how to satisfy water users. Even though countries nowadays consider options to reduce water demand in addition to options to increase supply, they generally do not include the global dimension of water management. In this way they do not explicitly consider options to save water through import of water-intensive products. In addition, by looking only at water use in their own country, most governments have a blind spot to the issue of sustainability of national consumption. As a matter of fact, many countries have significantly externalized their water footprint without looking whether the imported products are related to water depletion or pollution in the producing countries. Governments can and should engage with consumers and businesses to work towards sustainable consumer products. National water footprint accounting should be a standard component in national water statistics and provide a basis to formulate a national water plan and river basin plans that are coherent with national policies with respect to, for example, the environment, agriculture, industry, energy, trade, foreign affairs and international cooperation.

Water footprint and virtual water trade accounts can form a relevant input into the formulation of various sorts of governmental policy, such as: national or state water policy, river basin policy, local water policy, environmental policy, agricultural policy, industrial/economic policy, energy policy, trade policy, foreign policy and development cooperation policy (Table 5.5). Since the governmental organization can be regarded as a business in itself, another important thing for governments is to look at the possibility of reducing its own water footprint.

Key elements in a governmental strategy aimed at water footprint reduction are: raise water awareness among both consumers and producers, promote water-saving technology across all sectors of economy, restructure water-pricing mechanisms such that full costs of water-inputs become part of the cost of final commodities, promote product transparency throughout supply chains, and restructure economies based on unsustainable water supplies. These are all challenges that require inter-sector and in many cases also international cooperation. Political mandates are fragmented over different policy fields and levels, so the true assignment is to find out which measures need to be taken in different policy fields and at different levels to bring about the required concerted action.

Chapter 6

Limitations

The water footprint is a relatively new concept and water footprint assessment a new tool. As is often the case with new concepts and tools that are promising and speak to people's imaginations, expectations are not always realistic. Given the fact that the world's freshwater resources are limited, the water footprint is a very useful indicator, as it shows when, where and how consumers, producers and individual processes and products put a claim on this limited resource. Water footprint assessment is a useful tool to quantify and locate water footprints, to evaluate whether footprints are sustainable and to identify options to reduce water footprints where necessary. Having said that, the water footprint is no more than one relevant indicator in the very broad theme of sustainable, fair and efficient allocation and use of natural resources. Obviously it needs to be complemented with a wide array of other relevant indicators before integrated understanding can arise. Similarly, water footprint assessment is no more than one tool to understand the complex relations between societies and their environment. It focuses on the use of freshwater resources in the light of limited supplies. It does not address water issues that are not scarcity-related, such as flooding or lack of infrastructure for proper water supply to poor communities. It also does not address environmental issues other than freshwater scarcity.

Water footprint assessment is thus a partial tool, to be used in combination with other analytical means in order to provide understanding of the full range of relevant issues when making decisions. The rapid adoption of the water footprint as a comprehensive indicator of freshwater appropriation by human beings is very helpful in putting freshwater scarcity higher on the agenda's of governments and companies, but there is always the risk of oversimplification. There is a tendency in both governments and companies to reduce the complex reality into a very limited number of indicators. In governments, much focus is on the single indicator of 'gross national product' and in companies on the single indicator 'profit'. More widely spoken, governments generally focus on a limited number of social, environmental and economic indicators, where gross national product is one of the economic indicators. Companies generally use a limited number of 'key performance indicators', often categorized under the three terms 'people planet profit' (the triple bottom line as proposed by Elkington, 1997).

The water footprint can be seen as yet another indicator. Adding this indicator on the dashboard of policy-makers and chief executives is useful, but it suffers the same problem as all other widely-used environmental, social and economic indicators: it does not tell the whole story, but only reduces it into one simplistic measure. Indicators are only useful as long as they are used wisely.

Insights obtained from a water footprint analysis should always be combined with other relevant environmental, social, institutional, cultural, political and economic insights before well-informed decisions and related trade-offs can be made. Reducing and redistributing the water footprint of humanity is key in sustainable development, but other factors are key as well. Taking into account all other key factors is imperative when formulating strategies on how to apply different technical, institutional, political, communicative, economic and legal means to reduce water footprints.

In order to understand better what water footprint assessment is and is not, consider the following (non-exhaustive) list of limitations:

- Water footprint assessment focuses on analysing freshwater use in view of limited freshwater resources; it does not address other environmental themes such as climate change, depletion of minerals, fragmentation of habitats, limited land availability or soil degradation, nor does it address social or economic themes such as poverty, employment or welfare. Water footprint assessment touches on environmental, social and economic issues only insofar as the use of freshwater resources affects biodiversity, health, welfare or fair distribution. Obviously, when interested in the broader environmental, social and economic issues, much more factors play a role than freshwater alone. It is to be recognized that reducing humanity's water footprint where necessary is just one challenge, to be seen in a much broader context of other challenges.
- Water footprint assessment addresses the issues of freshwater scarcity and pollution. It does not address the issue of flooding. It also does not address the issue of people lacking access to proper clean water supply, since this is not a water scarcity issue but rather a poverty issue. Further, the water footprint is about freshwater, not about the use and pollution of seawater. Water footprint assessment is limited to considering those human activities that impact upon the quantity or quality of freshwater within a catchment or river basin.
- The water footprint is an indicator of freshwater use that considers consumptive water use and water pollution. This is interesting from a *catchment perspective*, since freshwater availability in a catchment is limited. The green, blue and grey water footprints show how human activities and products put a claim on these limited freshwater resources. Another useful

indicator of water use is the classical indicator of 'blue water withdrawal' (water abstraction). Knowing blue water abstractions is interesting as well, not so much from a catchment point of view, but from the water user point of view it is valuable to know all components of its water balance.

- Companies show an increasing interest in their 'water risk' (Levinson et al, 2008; Pegram et al, 2009; Morrison et al, 2009, 2010; Barton, 2010). Assessing the water footprint of a company helps to understand part of this risk, by showing which components in a company's water footprint are unsustainable, but a water footprint assessment is not the same as a full risk assessment. Unsustainable components in a company's water footprint assumingly imply a physical, reputational and regulatory risk for a company, affecting a company's social licence to operate, but if water risk is the focus of interest, carrying out a water footprint assessment is not sufficient.
- Governments have a broad responsibility in governing public resources. During the past few decades, it has been increasingly recognized that integrated approaches are important, in which consistency and coherence across different policy fields are seen as essential. In the field of water management, the integrated approach is generally known under the term 'integrated water resources management' (IWRM) and alternatively, when there is a specific catchment focus, under the term 'integrated river basin management' (IRBM) (GWP, 2000; GWP and INBO, 2009; UNESCO, 2009). IWRM and IRBM are very broad ideas, addressing both substantial questions such as 'what does a good integrated plan look like?' but also organizational questions such as 'how do we develop and implement such a plan?' and institutional questions such as 'how do we create proper enabling conditions?' The tool of 'water footprint assessment' is obviously not to replace IWRM or IRBM, but it should be regarded as an analytical tool that can help to broaden the knowledge base for IWRM and IRBM. Water footprint assessment broadens the traditional scope in water scarcity analysis by introducing supply chain thinking and including the international trade-related dimension of water scarcity and pollution. In this way, it can contribute to better-informed decisions being made in the context of water management.

Finally, it is worth mentioning that the water footprint concept, after having lived in the academic arena since 2002, did not enter the world of business, government and civil society until the second half of 2007. This means that experience with the concept in practice is limited. Therefore it is difficult to find many practical illustrations of full water footprint assessments. Most water footprint studies have thus far put emphasis on the accounting phase. Apart from global water footprint studies (Hoekstra and Chapagain, 2007a, 2008), a large amount of water footprint studies in a variety of geographical settings have been carried out (see an overview in Kuiper et al, 2010). The Spanish national government has been the first to formally embrace the water footprint concept by requiring the analysis of water footprints at river basin level in the preparation of river basin management plans (Official State Gazette, 2008; Garrido et al, 2010). A lot of companies have already analysed the water footprints of some of their products, but only a few companies have arrived at a stage where they can disclose some of the results (SABMiller and WWF-UK, 2009; SABMiller et al, 2010; TCCC and TNC, 2010; IFC et al, 2010; Chapagain and Orr, 2010). Few studies contain a full water footprint assessment as described in this manual. It is expected that, when more practical applications become available, this will provide valuable inputs for refining the procedures and methods as described here.

Chapter 7

Future Challenges

7.1 Water footprint assessment methodology and data

There are quite a number of practical issues that one will encounter when carrying out a water footprint assessment. In many cases this manual will give sufficient guidance, but in some cases there is an obvious need for further development of practical guidelines. A major question will often be how to handle the lack of required data. What default data should be used under such circumstances and what simplifications can be reasonably made? A major challenge is therefore to develop more detailed guidelines regarding what default data can be used when accurate local estimates are not available. In this context it is relevant to develop a database with default water footprint estimates for a large variety of processes and products, differentiating between production regions (such as countries). This would be very helpful for assessing the water footprints of consumers or producers, who know what they buy but often do not know all relevant details on the production and supply chain of the things they buy.

A practical issue in water footprint accounting is the truncation problem, which was already discussed in Section 2.2. The question here is what should be included and what can be excluded from the analysis. By applying a very broad scope of analysis when estimating the water footprint of a specific product, one will discover that some ingredients will not significantly contribute to the overall water footprint of the product and a continued further tracing of the supply chain does not yield additional value. More practical experience with water footprint accounting for a variety of products is necessary in order to be able to develop practical guidelines on what can – as a rule – be excluded from a product water footprint analysis. And also what consumer products or input products can be excluded from a consumer or business water footprint analysis, respectively.

An issue that has not yet received sufficient attention is how to handle variability and change in time. Not all, but many sorts of water use vary over the years, think for instance about the use of irrigation water in agriculture that depends on the rainfall pattern in a specific year (Garrido et al, 2010). Besides, water productivity may vary from year to year, due to all sorts of factors (including factors that have nothing to do with water), resulting in a variability of the water footprint over the years. Obviously, in this way, changes in a water footprint from one year to another cannot simply be interpreted as a structural improvement or worsening in water use. For that reason, water footprint data will often show a more meaningful picture if they show averages over a period of years. The question arises, what can best be taken as a period of analysis: five years, ten or even more? When will it be possible to analyse a trend in time? Besides, it will possibly appear that some sorts of input data can be best taken over a very long period (for example, 30 years as is usual for climate data), while other sorts of data can be taken per year or as an average over five years only. It would be useful to develop guidelines in this respect, acknowledging that, in the end, choices will also depend on the purpose of an analysis.

Related to the issue of variability, but even broader, is the issue of uncertainties. The uncertainties in data used in water footprint accounting can be very significant, which means that outcomes should be carefully interpreted. Carrying out an uncertainty analysis is definitely advisable, but often time restrictions will not allow for a very advanced uncertainty and sensitivity analysis. It would be useful to have at least some rough indications that show the order of magnitude of uncertainties in various sorts of water footprint accounts, so that one could refer to that. Currently, no uncertainty studies are available.

In terms of detail in water footprint accounting, one may find the distinction between a green, blue and grey water footprint too coarse. If desired, one can therefore split up blue water footprint accounts, for example, into surface water footprint, renewable-groundwater footprint and fossil-groundwater footprint accounts (see Section 3.3.1). The grey water footprint can be split up into pollutant-specific grey water footprint accounts (see Section 3.3.3).

In case of the grey water footprint, a challenge is to develop guidelines on how to define natural and maximum allowable concentrations. Both should ideally be catchment-specific, but in many cases such data are not available. Guidelines could advise to use a zero natural background for a specified list of chemicals and recommend which assumptions to make in case of other chemicals when catchment-specific values are not available. Besides, an issue that needs to be cleared up is whether one should take, for example, daily or monthly average concentrations. Maximum allowable concentrations for ambient water quality are not available for all substances; in these cases, guidelines should be available to advise what default values can best be used.

A question when measuring the blue water footprint is what resolution and scale can best be applied. What to do when water is withdrawn from one place and returned to another place downstream? According to the definition, the blue water footprint refers to 'consumptive water use', which refers to evapotranspiration, incorporation into a product, or to water that does not return to the same catchment area from which it was withdrawn. It obviously depends on the scale of analysis whether a return flow downstream of the withdrawal is consumptive or not. There may be cases of doubt, where very locally the water is regarded as consumptive but where at a larger scale it returns and thus is non-consumptive. Where to draw the line is something that needs to be found out in due course of time when more studies have been done and a good argument for a certain best scale can be made. Another question is what to do when groundwater is withdrawn and after use returned to fresh surface water. When blue water – referring to both groundwater and surface water – is considered as one category, this sort of interference will not be reflected in the blue water footprint. This is not a problem for many purposes, but in more detailed studies it may be desirable to distinguish a blue groundwater footprint and a blue surface water footprint. Further, in the case of groundwater, there is a crucial difference between renewable and fossil groundwater.

An interesting development is the use of remote sensing to estimate green and blue water footprints in agriculture at a high spatial and temporal resolution (Zwart et al, 2010; Romaguera et al, 2010), but further research is necessary to validate this approach and make it operational.

More research is needed on the quantification of catchment-specific 'environmental flow requirements' (Appendix V) and 'environmental green water requirements' (Box 4.3), because such data are an essential ingredient when assessing the sustainability of catchment-specific blue and green water footprints, respectively. Also, more research is required on the context-specific quantification of the maximum allowable decline of groundwater and lake levels (Box 4.5).

The chapter on water footprint sustainability assessment has shown that the definition of sustainability criteria deserves more attention, particularly the criteria for social and economic sustainability (Sections 4.2.3–4.2.4). Furthermore, it has been shown that the investigation of primary and secondary impacts is strongly dependent on choices made with respect to what sort of impacts to include and which ones to exclude from the assessment. This current manual provides little guidance on what impacts should be considered at least and which ones may be of less importance. It may be desirable to develop more guidance on what sorts of impacts to include depending on the purpose of the analysis.

Finally, it is necessary to develop understanding regarding how different sorts of policy response can contribute to the reduction of the green, blue and grey water footprints of various activities, including insight in the effectiveness of the different forms of response.

7.2 Application of the water footprint in different contexts

The number of applications of the water footprint concept is rapidly increasing. As one can see from the overview provided in Table 7.1, most studies have been published from 2007 onwards. The various water footprint studies that have been carried out thus far can be categorized into: global studies, national studies, regional and river basin studies, general product studies and company studies. Few studies address all phases of water footprint assessment, most studies have a strong or complete focus on water footprint accounting. The major challenge in future studies is to address the sustainability assessment and response formulation phases as well.

Table 7.1 *An overview of water footprint studies*

Global and supra-national water footprint and virtual water trade studies	• Global (Hoekstra and Hung, 2002, 2005; Hoekstra, 2003, 2006, 2008b; Chapagain and Hoekstra, 2004, 2008; Hoekstra and Chapagain, 2007a, 2008; Liu et al, 2009; Siebert and Döll, 2010) • Central Asia (Aldaya et al, 2010c)
National water footprint and virtual water trade studies	• China (Ma et al, 2006; Liu and Savenije, 2008; Hubacek et al, 2009; Zhao et al, 2009) • Germany (Sonnenberg et al, 2009) • India (Kumar and Jain, 2007; Kampman et al, 2008; Verma et al, 2009) • Indonesia (Bulsink et al, 2010) • Morocco (Hoekstra and Chapagain, 2007b) • Netherlands (Hoekstra and Chapagain, 2007b; Van Oel et al, 2008, 2009) • Romania (Ene and Teodosiu, 2009) • Spain (Novo et al, 2009; Aldaya et al, 2010b; Garrido et al, 2010) • Tunisia (Chahed et al, 2008) • UK (Chapagain and Orr, 2008; Yu et al, 2010)
Sub-national water footprint and virtual water trade studies	• Chinese provinces (Ma et al, 2006) • City of Beijing (Wang and Wang, 2009) • Indian states (Kampman et al, 2008) • Mancha Occidental Region, Spain (Aldaya et al, 2010d) • Andalusia, Spain (Dietzenbacher and Velazquez, 2007) • West Bank, Palestine (Nazer et al, 2008) • Guadiana basin, Spain (Aldaya and Llamas, 2008) • Lower Fraser Valley and the Okanagan basins, Canada (Brown et al, 2009) • Nile basin, Africa (Zeitoun et al, 2010)
Product water footprint studies	• bio-energy (Gerbens-Leenes et al, 2009a, 2009b; Gerbens-Leenes and Hoekstra, 2009, 2010; Dominguez-Faus et al, 2009; Yang et al, 2009; Galan-del-Castillo and Velazquez, 2010; Van Lienden et al, 2010) • coffee (Chapagain and Hoekstra, 2007; Humbert et al, 2009)

	• cotton (Chapagain et al, 2006b) • flowers (Mekonnen and Hoekstra, 2010b) • jatropha (Jongschaap et al, 2009; Maes et al, 2009; Gerbens-Leenes et al, 2009c; Hoekstra et al, 2009c) • mango (Ridoutt et al, 2010) • maize (Aldaya et al, 2010a) • meat (Chapagain and Hoekstra, 2003; Galloway et al, 2007; Hoekstra, 2010b) • onions (IFC et al, 2010) • paper (Van Oel and Hoekstra, 2010) • pasta (Aldaya and Hoekstra, 2010) • pizza (Aldaya and Hoekstra, 2010) • rice (Chapagain and Hoekstra, 2010) • soft drinks (Ercin et al, 2009) • soybean (Aldaya et al, 2010a) • sugar (Gerbens-Leenes and Hoekstra, 2009) • tea (Chapagain and Hoekstra, 2007) • tomatoes (Chapagain and Orr, 2009) • wheat (Liu et al, 2007; Aldaya et al, 2010a; Zwart et al, 2010; Mekonnen and Hoekstra, 2010a) • food in general (Chapagain and Hoekstra, 2004; Hoekstra and Chapagain, 2008; Hoekstra, 2008c)
Business water footprint studies	• beer from SABMiller (SABMiller and WWF-UK, 2009; SABMiller et al, 2010) • cola and orange juice from the Coca-Cola Company (TCCC and TNC, 2010) • breakfast cereal from Nestlé (Chapagain and Orr, 2010) • candies and pasta sauce from Mars (Ridoutt et al, 2009)

7.3 Embedding the water footprint in existing water and environmental accounts and reports

Traditional statistics on water use – whether national or corporate accounts – are mostly restricted to water withdrawals. The information basis is very narrow in this way, because it ignores green and grey water use and disregards indirect use as well. In the case of business accounts, the traditional approach pays no attention to water consumption and pollution in the supply chain. In the case of national accounts, the conventional approach overlooks virtual water imports and exports and the fact that part of the water footprint of national consumption lies outside the country. It will be necessary to gradually start incorporating water footprint statistics in governmental statistics and have them feature also in international statistics such as made available through, for example, the UN's Food and Agriculture Organization (FAO – AQUASTAT, FAOSTAT), the United Nations Environment Programme (UNEP – Geo Data Portal), the United Nations Development Programme (UNDP), the United

Nations Conference on Trade and Development (UNCTAD), the UN Statistics Division, the European Commission (Eurostat) and the World Bank. National water footprint statistics were already included in a number of international state-of-the-world publications (WWF, 2008, 2010; UN, 2010a). In the case of companies, it will be needed to start incorporating water footprint accounts in corporate environmental and sustainability reporting.

7.4 Linking to ecological, energy and carbon footprint methods

The water footprint is part of a family of footprint concepts. The oldest footprint concept is the ecological footprint, introduced in the 1990s by William Rees and Mathis Wackernagel (Rees, 1992; 1996; Rees and Wackernagel, 1994; Wackernagel and Rees, 1996). The ecological footprint measures the use of available bioproductive space and is measured in hectares. The carbon footprint concept originates from the ecological footprint discussion and has started to become more widely known since 2005 (Safire, 2008). The carbon footprint refers to the sum of greenhouse gas (GHG) emissions caused by an organization, event or product and is expressed in terms of CO_2 equivalents. Although the carbon footprint concept is relatively young, the idea of accounting GHG emissions is already much older; the first assessment of the Intergovernmental Panel on Climate Change (IPCC), for example, already dates back to 1990. Older than the ecological and carbon footprint concepts are also the concepts of 'embodied energy' and 'emergy' as applied in energy studies (Odum, 1996; Herendeen, 2004). These concepts refer to the total energy used to produce a product and are expressed in Joules.

The water footprint was introduced in the field of water studies in 2002 (Hoekstra, 2003). The term was chosen by analogy with the ecological footprint concept, but the roots of the water footprint are in water studies rather than environmental studies. Although the concepts of ecological footprints, water footprints, carbon footprints and embodied energy are thus very much related concepts, each of them has its own specific roots. As a result, the methods to quantify the different indicators show both striking similarities and differences. Two differences between the ecological and the water footprint are, for example, that ecological footprints are usually calculated based on global average productivities, while water footprints are calculated based on local productivities, and that ecological footprints are often not made spatially explicit, while water footprints are (Hoekstra, 2009).

The various 'footprint' concepts are to be regarded as complementary indicators of natural capital use in relation to human consumption. None of the indicators can substitute another, simply because each one provides another piece of information. Looking at only area requirements or only water or energy requirements is insufficient, since available land can be a critical factor in development, but so can available freshwater and energy. A challenge for future research is to bring the various footprint concepts and related methods together in one consistent conceptual and analytical framework.

7.5 Linking to material flow analysis, input-output modelling and life cycle assessment

Material flow analysis (MFA) is a method of analyzing the flows of materials in a well-defined system. On a national or regional scale, MFA can be used to study the material exchanges within an economy and between an economy and the natural environment. In industries, MFA can be used to analyse the material flows within a company or along an industrial supply chain involving a number of companies. When applied to a specific product, MFA refers to the study of inputs (resources) and outputs (emissions) along the different steps in the production system of a product. The latter sort of material flow analysis is similar to what is called the 'inventory phase' of life cycle assessment (LCA). LCA is the investigation and evaluation of the environmental impacts of a given product or service and consists of four phases: goal and scope, life cycle inventory, life cycle impact assessment and interpretation (Rebitzer et al, 2004).

Frameworks such as MFA, LCA and input-output modelling consider the use of various types of environmental resources and look at the various types of impacts on the environment. In contrast, ecological footprint, water footprint, carbon footprint and embodied energy analyses take the perspective of one particular resource or impact. Although it seems logical that 'footprints' are precisely the indicators typically used in MFA, LCA and input-output studies, the methods applied in footprint studies and the methods applied in MFA, LCA and input-output studies do not form one coherent framework of methods. Hitherto, from a water perspective, MFA, LCA and input-output studies do not include freshwater in a sufficient way.

In the input-output research community, there is an increasing interest to include water (see, for example, Dietzenbacher and Velazquez, 2007; Zhao et al, 2009; Wang and Wang, 2009; Yu et al, 2010). Water footprint studies can provide data on the operational water footprint per economic sector, which are necessary input data in environmentally extended input-output studies that aim to include freshwater use.

Also within the LCA community there is an increasing interest in water (Koehler, 2008; Milà i Canals et al, 2009). LCA studies aim to assess the overall

Table 7.2 *How water footprint assessments can feed LCA*

Water footprint assessment phase	*Outcome*	*Physical meaning*	*Resolution*	*LCA phase*
Product water footprint accounting (Section 3.4)	Green, blue and grey water footprints (volumetric)	Water volume consumed or polluted per unit of product	Spatiotemporally explicit	Life cycle inventory
Product water footprint sustainability assessment (Section 4.4.1)	An evaluation of the sustainability of a green, blue and grey product water footprint from an environmental, social and economic perspective	Various measurable impact variables	Spatiotemporally explicit	Life cycle impact assessment
Aggregation of selected information from the water footprint sustainability assessment (Section 4.4.2)	Aggregated water footprint impact indices	None	Non spatiotemporally explicit	

Source: Based on Hoekstra et al (2009b)

environmental impact of products. The use of freshwater has, until recently, received insufficient attention in LCA studies. There are two separate issues. First, freshwater resources in the world are limited, so measuring freshwater appropriation by looking at water consumption and pollution in terms of volumes should therefore be a key element in an LCA study. The green, blue and grey water footprints of a product are good indicators for this total freshwater appropriation. Second, one can look at the *local environmental impacts* related to the freshwater appropriation. For this latter purpose, water footprint accounting and sustainability assessment can serve LCA studies as summarized in Table 7.2. A product water footprint account contributes to the life cycle inventory for the product; the water footprint sustainability assessment contributes to the life cycle impact assessment.

Some LCA authors have suggested using the term 'water footprint' for what in this manual is called the 'blue water footprint impact index' (Section 4.4.2). In this way, the term 'water footprint' would no longer be a volumetric measure of freshwater appropriation, but a local environmental impact index (Pfister et al, 2009; Ridoutt et al, 2009; Ridoutt and Pfister, 2010; Berger and Finkbeiner, 2010). It has also been proposed to neglect green water footprints, because impacts would be nil (Pfister and Hellweg, 2009). Redefining the water footprint does not make sense, however, from a water resources management perspective, which requires spatially and temporally explicit information on water footprints in real volumes and impacts in real terms as well. Water footprint studies serve two discourses in water resources management. First, data on water footprints of products, consumers, and producers inform the discourse about sustainable, equitable and efficient freshwater use and allocation. Freshwater is scarce, its annual availability is limited. It is relevant to know who receives which portion and how water is allocated over various purposes. For example, rainwater used for bio-energy cannot be utilized for food. Second, water footprint accounts help to estimate environmental, social and economic impacts at catchment level. Environmental impact assessment should include a comparison of each water footprint component to available water at relevant locations and time (accounting for environmental water requirements). The proposal to use the term water footprint for what is called here the 'blue water footprint impact index' is highly confusing. It has been proposed that the water footprint should include environmental impact because the carbon footprint would also do that, namely by weighing different greenhouse gases and expressing them in terms of CO_2 equivalents. The fact that greenhouse gases are measured in terms of CO_2 equivalents, however, does not mean that the carbon footprint reflects environmental impacts of greenhouse gases. The carbon footprint is a measure of how much greenhouse gases are emitted into the environment caused by human activities. It does not show anything one can reasonably call 'impact'. It just measures emissions, brought into one common denominator. The carbon footprint does not describe the environmental impact of greenhouse gases, such as increased temperatures and changed patterns of evaporation and precipitation. In this sense, the water footprint, carbon footprint and ecological footprint are similar concepts. The carbon footprint measures total emission of greenhouse gases; the ecological footprint measures the total use of bioproductive space; the water footprint measures the total appropriation of freshwater. The total use of space or water does not reflect impacts, in the same way as the total emission of greenhouse gases does not reflect impacts. Footprints show the pressure of humans on the environment, not the impacts.

Chapter 8

Conclusion

This manual contains the global standard for the definition of water footprint terms and the method of water footprint assessment. The standard has been developed over the past two years in an open and transparent process facilitated by the Water Footprint Network. Through the involvement of a large variety of organizations around the world that thus far have been working with the water footprint concept, the standard has become a true globally shared standard. Having a unique and shared standard is increasingly important since communicating about water footprints and desired water footprint reductions will be seriously hampered if different stakeholders use different ways of defining and calculating water footprints.

It is envisaged that future developments will allow refinements of the standard and particularly the development of practical guidelines. The method of water footprint accounting (Chapter 3) is now, after eight years of continued development, firmly established and widely adopted, both in the scientific community and in practice. The increased use of the water footprint in practical contexts in the past few years has contributed to maturing of the concept. Nevertheless, various challenges remain, including the development of practical guidelines per product category and business sector on how to truncate the analysis (where to stop going back along supply chains) and rules on how to account for uncertainties and how to deal with time variability when doing trend analysis. Besides, there is a huge challenge to develop databases on typical process water footprints (the basic ingredient for each analysis) and software tools to make it easier for practitioners to set up a water footprint account. Following the guidelines on water footprint accounting as provided in this manual is much more labour-intensive than when one could use a simple computer tool guiding the analysis. Developing such a tool together with underlying databases is therefore part of the work programme of the Water Footprint Network (WFN).

The chapters on water footprint sustainability assessment and response options (Chapters 4 and 5) are less mature than the chapter on water footprint accounting (Chapter 3). This is due to the fact that these two phases of water footprint assessment have got less attention so far, both in scientific studies and in practical implementation. The chapter on water footprint sustainability assessment is limited to a description of the procedure of sustainability assessment and a discussion of the major sustainability criteria to be considered. The chapter on response options is mainly an inventory of responses that could be considered. In this respect, the manual offers a reference framework for analysing sustainability and response options rather than an in-depth treatment of how to carry out a full impact assessment or how specific response options can be studied in more detail in terms of their implications and strengths and weaknesses. In addition, it should be noted that the chapters on sustainability assessment and response should not be read as recipes that lead to final answers on what to do. Although a recipe may sound attractive to some – particularly in the daily practice of companies it would be helpful to have clear, unambiguous regulations – the reality is that assessing sustainability and formulating a course of response are activities that contain numerous subjective, value-laden elements. The ambition of the chapters on water footprint sustainability assessment and policy response options is to offer a rough guideline, not a detailed recipe.

The broad interest in the water footprint concept and methodology has taken off in September 2007 with a small meeting between representatives from civil society, business, academia and the UN. Since then, the interest in applying the water footprint in governmental policy and corporate strategy has been growing continuously. This has led to the establishment of the WFN on 16 October 2008. Twelve months later, the network had 76 partners, coming from all continents and from all sorts of sectors: government, business, investors, civil society, intergovernmental institutions, consultants, universities and research institutes. At the time of the finalization of the manuscript of this manual, 16 October 2010, precisely two years after the establishment of the WFN, the network had 130 partners. A major challenge is to maintain a shared language in the field of water footprint assessment, because concrete targets towards sustainable water resources use can only be transparent, meaningful and effective when formulated in a common terminology and based on a shared calculation methodology. This water footprint assessment manual provides such a common base. Adjustments and refinement to the manual will be made in the future based on new research and development and on experiences from practitioners working with the method in their own practice.

Appendix I

Calculation of Green and Blue Evapotranspiration Using the CROPWAT Model

The 'CWR option' in the CROPWAT model

Green and blue water evapotranspiration during crop growth can be estimated with the Food and Agriculture Organization's CROPWAT model (FAO, 2010b). The model offers two alternative options. The simplest but not the most accurate option is the 'CWR option'. In this option, it is assumed that there are no water limitations to crop growth. The model calculates: (i) crop water requirements (*CWR*) during the full length of the growing period under particular climatic circumstances; (ii) effective precipitation over the same period; (iii) irrigation requirements.

The crop water requirement is the water needed for evapotranspiration under ideal growth conditions, measured from planting to harvest. 'Ideal conditions' means that adequate soil water is maintained by rainfall and/or irrigation so that it does not limit plant growth and crop yield. Basically, the crop water requirement is calculated by multiplying the reference crop evapotranspiration (ET_0) by the crop coefficient (K_c): $CWR = K_c \times ET_0$. It is assumed that the crop water requirements are fully met, so that actual crop evapotranspiration (ET_c) will be equal to the crop water requirement: $ET_c = CWR$.

The reference crop evapotranspiration ET_0 is the evapotranspiration rate from a reference surface, not short of water. The reference crop is a hypothetical surface with extensive green grass cover with specific standard characteristics and therefore the only factors affecting ET_0 are climatic parameters. ET_0 expresses the evaporating power of the atmosphere at a specific location and time of the year and does not consider the crop characteristics and soil factors. The actual crop evapotranspiration under ideal conditions differs distinctly from the reference crop evapotranspiration, as the ground cover, canopy properties and aerodynamic resistance of the crop are different from the grass used as reference. The effects of characteristics that distinguish field crops from grass are integrated into the crop coefficient (K_c). The crop coefficient varies over the length of the growing period. Values for K_c for different crops over the length of the growing period can be taken from the literature (for example, Allen et al, 1998). As an alternative, one can calculate K_c as the sum of K_{cb} and K_e, where K_{cb} is the so-called basal crop coefficient and K_e a soil evaporation coefficient. The basal crop coefficient is defined as the ratio of the crop evapotranspiration over the reference evapotranspiration (ET_c/ET_o) when the soil surface is dry but transpiration is occurring at a potential rate, in other words, water is not limiting transpiration. Therefore, $K_{cb} \times ET_o$ represents primarily the transpiration component of ET_c, but it also includes a residual diffusive evaporation component supplied by soil water below the dry surface and by soil water from beneath dense vegetation. The soil evaporation coefficient K_e describes the evaporation component of ET_c. When the topsoil is wet, following rain or irrigation, K_e is maximal; when the soil surface is dry, K_e is small and even zero when no water remains near the soil surface for evaporation. Different irrigation techniques wet the soil surface in different degrees. Sprinkler irrigation, for example, wets the soil more than drip irrigation, resulting in a higher value for K_e directly after irrigation. This will translate into a higher value for K_c and thus for ET_c. The CROPWAT model, however, does not allow the specification of K_{cb} and K_e separately; it requires specification of the resultant K_c. Besides, K_c cannot be specified per day but only for three different periods in the growing period, so that the effect of different irrigation techniques can be simulated in CROPWAT only by roughly adjusting K_c as a function of the irrigation technique used. On average, K_c will be higher when irrigation techniques are applied that wet the soil intensively than when techniques are used that do not wet the top soil much. As an alternative to CROPWAT, one can decide to use AQUACROP (FAO, 2010e), a crop model that better simulates crop yield under water stress conditions and that separates K_{cb} and K_e.

Effective precipitation (P_{eff}) is the part of the total amount of precipitation that is retained by the soil so that it is potentially available for meeting the water need of the crop. It is often less than the total rainfall because not all rainfall can actually be appropriated by the crop, for example, due to surface run-off or percolation (Dastane, 1978). There are various ways to estimate effective rainfall based on total rainfall; Smith (1992) recommends the USDA SCS method (the method of the United States Department of Agriculture, Soil Conservation Service). This is one of the four alternative methods that the users of CROPWAT can choose from.

The irrigation requirement (IR) is calculated as the difference between crop water requirement and effective precipitation. The irrigation requirement is zero if effective rainfall is larger than the crop water requirement. This means: $IR = \max(0, CWR - P_{eff})$. It is assumed that the irrigation requirements are fully met. Green water evapotranspiration (ET_{green}), in other words, evapotranspiration of rainfall, can be equated with the minimum of total crop evapotranspiration (ET_c) and effective rainfall (P_{eff}). Blue water evapotranspiration (ET_{blue}), in other words, field-evapotranspiration of irrigation water, is equal to the total crop evapotranspiration minus effective rainfall (P_{eff}), but zero when effective rainfall exceeds crop evapotranspiration:

$$ET_{green} = \min(ET_c, P_{eff}) \quad \text{[length/time]} \tag{59}$$

$$ET_{blue} = \max(0, ET_c - P_{eff}) \quad \text{[length/time]} \tag{60}$$

All water flows are expressed in mm/day or in mm per period of simulation (e.g. ten days).

The 'irrigation schedule option' in the CROPWAT model

Green and blue water evapotranspiration during crop growth can be estimated with FAO's CROPWAT model (FAO, 2010b). The model offers two alternative options. The 'irrigation schedule option' is more accurate and not much more complex than the 'CWR option', allowing the specification of the actual irrigation over the growing period. The model does not work with the concept of effective precipitation (as in the case of the 'CWR option', see above). Instead, the model includes a soil water balance which keeps track of the soil moisture content over time using a daily time step. For this reason, the model requires input data on soil type. The calculated evapotranspiration is called ET_a, the adjusted crop evapotranspiration, which may be smaller than ET_c due to non-optimal conditions. ET_a is calculated as the crop evapotranspiration under optimal conditions (ET_c) times a water stress coefficient (K_s):

$$ET_a = K_s \times ET_c = K_s \times K_c \times ET_o \quad \text{[length/time]} \tag{61}$$

The stress coefficient K_s describes the effect of water stress on crop transpiration. For soil water limiting conditions, $K_s < 1$; when there is no soil water stress, $K_s = 1$. For the crop coefficient K_c the same can be said as what has been said already for the 'CWR option' above.

Rain-fed conditions can be simulated by the model by choosing to apply no irrigation. In the rain-fed scenario, the green water evapotranspiration (ET_{green}) is equal to the total evapotranspiration as simulated by the model and the blue water evapotranspiration (ET_{blue}) is zero.

Irrigated conditions can be simulated by specifying how the crop is irrigated. Different irrigation timing and application options can be selected depending on the actual irrigation strategy. The default option, 'irrigate at critical depletion' and 'refill soil to field capacity', assumes 'optimal' irrigation where the irrigation intervals are at a maximum while avoiding any crop stress. The average irrigation application depth per irrigation period is related to the irrigation method practised. Generally, in the case of high frequency irrigation systems, such as micro-irrigation and centre pivot, about 10mm or less per wetting event are applied. In the case of surface or sprinkler irrigation, irrigation depths of 40mm or more are common. After running the model with the selected irrigation options, the total water evapotranspired (ET_a) over the growing period is equal to what is called 'actual water use by crop' in the model output. The blue water evapotranspired (ET_{blue}) is equal to the minimum of 'total net irrigation' and 'actual irrigation requirement' as specified in the model output. The green water evapotranspired (ET_{green}) is equal to the total water evapotranspired (ET_a) minus the blue water evapotranspired (ET_{blue}) as simulated in the irrigation scenario.

Alternatively, one can run two scenarios: with and without irrigation. In both scenarios one should take the crop characteristics (such as rooting depth) as they are under conditions of irrigation, because these characteristics may differ considerably for irrigated and rain-fed farming. The green water evapotranspiration under irrigation conditions can be estimated by assuming it to be equal to the total evapotranspiration as simulated in the scenario without irrigation. The blue water evapotranspiration can be calculated as the total evapotranspiration as simulated in the scenario with irrigation minus the estimated green water evapotranspiration.

Note that, over the growing period as a whole, blue water evapotranspiration is generally less than the actual irrigation volume applied. The difference refers to irrigation water that percolates to the groundwater or runs off from the field.

Appendix II

Calculating the Process Water Footprint of Growing a Crop: An Example for Sugar Beet in Valladolid (Spain)

This appendix provides an example of how to estimate the green, blue and grey process water footprints of growing a crop. It focuses on the case of a sugar beet (*Beta vulgaris* var. *vulgaris*) production in a one-hectare irrigated crop field in Valladolid (north-central Spain).

Green and blue components of the process water footprint

First, the green-blue water evapotranspiration has been estimated using the CROPWAT 8.0 model (Allen et al, 1998; FAO, 2010b). There are two different ways to do this: using the crop water requirement option (assuming optimal conditions) or the irrigation schedule option (including the possibility to specify actual irrigation supply in time). A comprehensive manual for the practical use of the program is available online (FAO, 2010b).

In both cases, the calculations have been done using climate data from the nearest and most representative meteorological station located in the crop-producing region (Figure II.1). When possible, crop data were obtained from local agricultural research stations. The planting dates at provincial level were obtained from the Spanish Ministry of Agriculture, Fisheries and Food (MAPA, 2001) (Table II.1). In the temperate north of Spain, beets are planted in the spring and harvested in the autumn. In warmer southern areas (Andalusia), sugar beets are a winter crop, planted in the autumn and harvested in the spring. Crop coefficients and crop lengths according to the type of region and climate were taken from the UN's Food and Agriculture Organization (FAO) (Allen et al, 1998, Tables 11 and 12). Data on rooting depth, critical depletion level and yield response factor were obtained from FAO global databases (FAO, 2010b). Besides, in the irrigation schedule option, soil data are required to estimate the soil water balance. Soil information was also obtained from FAO (2010b).

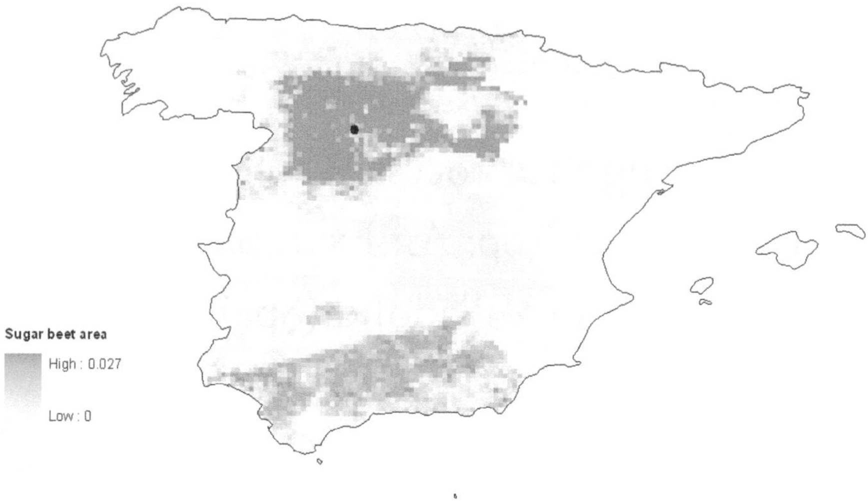

Figure II.1 Climate station in Valladolid (Spain) (dot in black) and sugar beet harvested area in Spain (unit: proportion of grid cell area)

Source of sugar beet area: Monfreda et al (2008)

Table II.1 *Planting and harvesting dates and yield for sugar beet production in Valladolid (Spain)*

Crop	*Planting date**	*Harvesting date**	*Yield (ton/ha)***
Sugar beet	1 April (March-April)	27 Sept (Sept-Oct)	81

* Source: MAPA (2001)
** Source: MARM (2009) period 2000–2006

Crop water requirement option

This option estimates evapotranspiration under optimal conditions, which means that crop evapotranspiration (ET_c) equals the crop water requirement (CWR). Optimal means disease-free, well-fertilized crops, grown in large fields, under optimum soil water conditions and achieving full production under the given climatic conditions (Allen et al, 1998). The crop water requirement option can be run with climate and crop data alone. ET_c is estimated with a ten day time step and over the total growing season using the effective rainfall. To calculate the effective rainfall, the method of the Soil Conservation Service of the United States Department of Agriculture (USDA SCS) was chosen as it is one of the most widely used methods. The model calculates ET_c as follows:

Table II.2 *Total green-blue water evapotranspiration based on the CWR output table of CROPWAT 8.0*

Month	Period	Stage	K_c –	ET_c mm/day	ET_c mm/period	P_{eff} mm/period	Irr. req. mm/period	ET_{green} mm/period	ET_{blue} mm/period
Apr	1	Init	0.35	1.02	10.2	12.6	0	10.2	0
Apr	2	Init	0.35	1.13	11.3	13.8	0	11.3	0
Apr	3	Init	0.35	1.24	12.4	14	0	12.4	0
May	1	Init	0.35	1.35	13.5	14.5	0	13.5	0
May	2	Init	0.35	1.45	14.5	15	0	14.5	0
May	3	Dev	0.48	2.2	24.2	13.8	10.4	13.8	10.4
Jun	1	Dev	0.71	3.55	35.5	12.7	22.7	12.7	22.8
Jun	2	Dev	0.94	5.02	50.2	11.9	38.3	11.9	38.3
Jun	3	Mid	1.15	6.6	66	9.8	56.3	9.8	56.2
Jul	1	Mid	1.23	7.58	75.8	7.1	68.6	7.1	68.7
Jul	2	Mid	1.23	8.05	80.5	5	75.6	5	75.5
Jul	3	Mid	1.23	7.8	85.8	4.8	81	4.8	81
Aug	1	Mid	1.23	7.59	75.9	4.1	71.8	4.1	71.8
Aug	2	Late	1.23	7.39	73.9	3.3	70.6	3.3	70.6
Aug	3	Late	1.13	6.05	66.6	5.7	60.9	5.7	60.9
Sep	1	Late	1	4.65	46.5	8.9	37.5	8.9	37.6
Sep	2	Late	0.87	3.51	35.1	11.2	23.8	11.2	23.9
Sep	3	Late	0.76	2.6	18.2	7.8	7	7.8	10.4
Over the total growing period					796	176	625	168	628

$$ET_c = K_c \times ET_o \quad \text{[length/time]} \tag{62}$$

Here, K_c refers to the crop coefficient, which incorporates crop characteristics and averaged effects of evaporation from the soil. ET_o represents the reference evapotranspiration, which expresses the evapotranspiration from a hypothetical grass reference crop not short of water.

The green water evapotranspiration (ET_{green}) is calculated as the minimum of total crop evapotranspiration (ET_c) and effective rainfall (P_{eff}), with a time step of ten days. The total green water evapotranspiration is obtained by summing up ET_{green} over the growing period. The blue water evapotranspiration (ET_{blue}) is estimated as the difference between the total crop evapotranspiration (ET_c) and the total effective rainfall (P_{eff}) on a ten-day basis. When the effective rainfall is greater than the crop total crop evapotranspiration ET_{blue} is equal to zero. The total blue water evapotranspiration is obtained by adding ET_{blue} over the whole growing period (Table II.2).

$$ET_{green} = \min(ET_c, P_{eff}) \quad \text{[length/time]} \qquad (63)$$

$$ET_{blue} = \max(0, ET_c - P_{eff}) \quad \text{[length/time]} \qquad (64)$$

Irrigation schedule option

In the second option we can calculate the crop evapotranspiration under both optimal and non-optimal conditions over the total growing season using the daily soil water balance approach. The calculated evapotranspiration is called ET_a, the adjusted crop evapotranspiration. ET_a may be smaller than ET_c due to non-optimal conditions. The water movements in the soil, the water holding capacity of the soil and the ability of the plants to use the water can be influenced by different factors, such as physical condition, fertility and biological status of the soil. ET_a is calculated using a water stress coefficient (K_s):

$$ET_a = K_s \times ET_c = K_s \times K_c \times ET_o \quad \text{[length/time]} \qquad (65)$$

K_s describes the effect of water stress on crop transpiration. For soil water limiting conditions, $K_s < 1$; when there is no soil water stress, $K_s = 1$.

The irrigation schedule option requires climate, crop and soil data. To estimate the green water evapotranspiration (ET_{green}) in rain-fed agriculture, the 'no irrigation (rain-fed)' choice is selected within the 'options' button on the Toolbar (Table II.3). Under this scenario, the green water evapotranspired (ET_{green}) is equal to the total evapotranspiration as simulated, which is given under 'actual water use by crop' as specified in the model output. The blue water evapotranspired (ET_{blue}) is obviously zero in this case.

To estimate the green and blue water evapotranspiration in irrigated agriculture, different irrigation timing and application options can be selected depending on the actual irrigation strategy. The default option, 'irrigate at critical depletion' and 'refill soil to field capacity', assumes 'optimal' irrigation where the irrigation intervals are at a maximum while avoiding any crop stress. The average irrigation application depth per irrigation is related to the irrigation method practised. Generally, in the case of high frequency irrigation systems, such as micro-irrigation and centre pivot, about 10mm or less per wetting event are applied. In the case of surface or sprinkler irrigation, irrigation depths are 40mm or more. In the sugar beet production in Valladolid, 40mm are applied every seven days (Table II.4). After running the model with the selected irrigation options, the total water evapotranspired is equal to ET_a over the growing period as given in the model output ('actual water use by crop'). After running the

Table II.3 *Irrigation schedule under the rain-fed scenario: Output table of CROPWAT 8.0*

CROP IRRIGATION SCHEDULE

ETo station: VALLADOLID Crop: Sugar beet Planting date: 01/04
Rain station: VALLADOLID Soil: Medium (loam) Harvest date: 27/09

Yield red.: 50.1%

Crop scheduling options
Timing: No irrigation (rain-fed)
Application: –
Field eff. 70%

Table format: Daily soil moisture balance

Date	Day	Stage	Rain mm	K_s –	ET_a mm	Depl %	Net Irr mm	Deficit mm	Loss mm	Gr. Irr mm	Flow l/s/ha
01-Apr	1	Init	0	1	1	1	0	1	0	0	0
02-Apr	2	Init	0	1	1	2	0	2	0	0	0
03-Apr	3	Init	6.7	1	1	1	0	1	0	0	0
04-Apr	4	Init	0	1	1	2	0	2	0	0	0
05-Apr	5	Init	0	1	1	3	0	3	0	0	0
06-Apr	6	Init	0	1	1	4	0	4.1	0	0	0
07-Apr	7	Init	6.7	1	1	1	0	1	0	0	0
08-Apr	8	Init	0	1	1	2	0	2	0	0	0
09-Apr	9	Init	0	1	1	3	0	3	0	0	0
10-Apr	10	Init	0	1	1	4	0	4.1	0	0	0
11-Apr	11	Init	0	1	1.1	5	0	5.2	0	0	0
12-Apr	12	Init	0	1	1.1	6	0	6.3	0	0	0
13-Apr	13	Init	7.4	1	1.1	1	0	1.1	0	0	0
...											
25-Sep	178	End	0	0.21	0.5	92	0	266.5	0	0	0
26-Sep	179	End	0	0.2	0.5	92	0	267	0	0	0
27-Sep	End	End	0	0.2	0	90					

Totals:

Total gross irrigation	0	mm	Total rainfall	190.3	mm
Total net irrigation	0	mm	Effective rainfall	171.1	mm
Total irrigation losses	0	mm	Total rain loss	19.3	mm
Actual water use by crop	432.2	mm	Moist deficit at harvest	261.1	mm
Potential water use by crop	793.3	mm	Actual irrigation requirement	622.3	mm
Efficiency irrigation schedule	–	%	Efficiency rain	89.9	%
Deficiency irrigation schedule	45.5	%			

Yield reductions:

Stage label	A	B	C	D	Season	
Reductions in ET_c	0	0	53.3	87.7	45.5	%
Yield response factor	0.5	0.8	1.2	1	1.1	
Yield reduction	0	0	64	87.7	50.1	%
Cumulative yield reduction	0	0	64	95.6		%

Table II.4 *Irrigation schedule under the irrigation scenario: Output table of CROPWAT 8.0*

CROP IRRIGATION SCHEDULE

ETo station: VALLADOLID   Crop: Sugar beet   Planting date: 01/04
Rain station: VALLADOLID   Soil: Medium (loam)   Harvest date: 27/09

Yield red.: 0.0%

Crop scheduling options
Timing: Irrigate at user defined intervals
Application: Fixed application depth of 40mm
Field eff. 70%

Table format: Daily soil moisture balance

Date	Day	Stage	Rain mm	K_s –	ET_a mm	Depl %	Net Irr mm	Deficit mm	Loss mm	Gr. Irr mm	Flow l/s/ha
01-Apr	1	Init	0	1	1	1	0	1	0	0	0
02-Apr	2	Init	0	1	1	2	0	2	0	0	0
03-Apr	3	Init	6.7	1	1	1	0	1	0	0	0
04-Apr	4	Init	0	1	1	2	0	2	0	0	0
05-Apr	5	Init	0	1	1	3	0	3	0	0	0
06-Apr	6	Init	0	1	1	4	0	4.1	0	0	0
07-Apr	7	Init	6.7	1	1	1	40	0	39	57.1	6.61
08-Apr	8	Init	0	1	1	1	0	1	0	0	0
09-Apr	9	Init	0	1	1	2	0	2	0	0	0
10-Apr	10	Init	0	1	1	3	0	3	0	0	0
11-Apr	11	Init	0	1	1.1	4	0	4.2	0	0	0
12-Apr	12	Init	0	1	1.1	5	0	5.3	0	0	0
13-Apr	13	Init	7.4	1	1.1	1	0	1.1	0	0	0
…											
25-Sep	178	End	0	1	2.6	6	0	16.3	0	0	0
26-Sep	179	End	0	1	2.6	7	0	18.9	0	0	0
27-Sep	End	End	0	1	0	4					

Totals:

Total gross irrigation	1428.6	mm	Total rainfall	190.3	mm
Total net irrigation	1000.0	mm	Effective rainfall	125.1	mm
Total irrigation losses	344.8	mm	Total rain loss	65.2	mm
Actual water use by crop	793.3	mm	Moist deficit at harvest	13.0	mm
Potential water use by crop	793.3	mm	Actual irrigation requirement	668.3	mm
Efficiency irrigation schedule	65.5	%	Efficiency rain	65.7	%
Deficiency irrigation schedule	0.0	%			

Yield reductions:

Stage label	A	B	C	D	Season	
Reductions in ET_c	0	0	0	0	0	%
Yield response factor	0.5	0.8	1.2	1	1.1	
Yield reduction	0	0	0	0	0	%
Cumulative yield reduction	0	0	0	0		%

model with the selected irrigation options, the total water evapotranspired (ET_a) over the growing period is equal to what is called 'actual water use by crop' in the model output. The blue water evapotranspired (ET_{blue}) is equal to the minimum of 'total net irrigation' and 'actual irrigation requirement' as specified in the model output. The green water evapotranspired (ET_{green}) is equal to the total water evapotranspired (ET_a) minus the blue water evapotranspired (ET_{blue}) as simulated in the irrigation scenario.

In both options (CWR and irrigation schedule), the estimated crop evapotranspiration in mm is converted to m³/ha applying the factor 10. The green component in the process water footprint of a crop ($WF_{proc,green}$, m³/ton) is calculated as the green component in crop water use (CWU_{green}, m³/ha) divided by the crop yield Y (ton/ha). The blue component ($WF_{proc,blue}$, m³/ton) is calculated in a similar way:

$$WF_{proc,green} = \frac{CWU_{green}}{Y} \quad \text{[volume/mass]} \qquad (66)$$

$$WF_{proc,blue} = \frac{CWU_{blue}}{Y} \quad \text{[volume/mass]} \qquad (67)$$

The outcome of both options is given in Table II.5. The results are similar with respect to the total ET and the resultant total water footprint, but quite different with respect to the ratio blue/green.

The calculations above refer to the evapotranspiration from the field; we have not yet accounted for the green and blue water incorporated into the harvested crop. The water fraction of sugar beet is in the range of 75–80 per cent, which means that the water footprint of sugar beet is 0.75–0.80m³/ton if we look at incorporated water alone. This is less than 1 per cent of the water footprint related to evaporated water.

Table II.5 *Calculation of the green and blue components of the process water footprint (m³/ton) for sugar beet in Valladolid (Spain) using the CWR-option and irrigation schedule option for a medium soil*

CROPWAT option	ET_{green}	ET_{blue}	ET_a	CWU_{green}	CWU_{blue}	CWU_{tot}	Y*	$WF_{proc,green}$	$WF_{proc,blue}$	WF_{proc}
	mm / growing period			m³/ha			ton/ha	m³/ton		
Crop water requirement option	168	628	796	1680	6280	7960	81	21	78	98
Irrigation schedule option	125	668	793	1250	6680	7930	81	15	82	98

* Source: MARM (2009) period 2000–2006

Table II.6 *Calculation of the grey component of the process water footprint (m^3/ton) for sugar beet in Valladolid (Spain)*

*Average fertilizer application rate**	*Area*	*Total fertilizer applied*	*Nitrogen leaching or running off to water bodies 10%*	*max. conc.*	*Total $WF_{proc,grey}$ sugar beet*	*Production***	*$WF_{proc,grey}$ sugar beet*
kg/ha	ha	ton/year	ton/year	mg/l	10^6 m^3/year	ton	m^3/ton
178	1	0.2	0.02	10	0.002	81	22

* Source: FertiStat (FAO, 2010c)
** Source: MARM (2009) period 2000–2006

Grey component in the process water footprint

The grey component in the process water footprint of a primary crop (m^3/ton) is calculated as the load of pollutants that enters the water system (kg/yr) divided by the difference between the ambient water quality standard for that pollutant (the maximum acceptable concentration c_{max}) and its natural concentration in the receiving water body (c_{nat}) (Table II.6). The quantity of nitrogen that reaches free flowing water bodies has been assumed to be 10 per cent of the applied fertilization rate (in kg/ha/yr) (Hoekstra and Chapagain, 2008). The effect of the use of other nutrients, pesticides and herbicides to the environment has not been analyzed. The total volume of water required per ton of nitrogen is calculated considering the volume of nitrogen that leaches or runs off (ton/ton) and the maximum allowable concentration in the free flowing surface water bodies. As ambient water quality standard for nitrogen, we have used 10mg/litre (measured as N). This limit was used to calculate the volume of freshwater required to assimilate the load of pollutants. By lack of appropriate data, the natural concentration in the receiving water body was assumed to be zero. Data on the application of fertilizers have been obtained from the FertiStat database (FAO, 2010c).

Appendix III

Calculating the Water Footprint of a Product: Example for Refined Sugar from Valladolid (Spain)

This appendix provides an example of how to estimate the green, blue and grey water footprints of a product focusing on the case of the refined sugar production from Valladolid (Spain).

If a primary crop is processed into a crop product (for example, sugar beet processed into raw sugar), there is often a loss of weight, because only part of the primary product is used. The water footprint of crop products is calculated by dividing the water footprint of the input product by the product fraction. The product fraction is defined as the quantity of the output product obtained per quantity of input product. The product fractions for various crop products are derived from different commodity trees as defined in FAO (2003) and Chapagain and Hoekstra (2004). Figure III.1 gives the product tree for refined sugar. If the input product is processed into two or more different products, one needs to distribute the water footprint of the input product across its separate products. This is done proportionally to the value of the input products. The value fraction for a processed product is defined as the ratio of the market value of the output product to the aggregated market value of all the output products obtained from the input product. If during processing there is some water use involved, the process water use is added to the water footprint value of the root product before the total is distributed over the various processed products.

The sugar beet naturally contains sugar. In a sugar production plant, this sugar is removed from the beet and converted into granulated sugar. The beet harvest begins in mid-September. Most of the beet is delivered by truck. The beet delivered is first washed in water in large washing units. The water used is cleaned in the water purification plant for reuse. The soil that is removed is first stored in storage fields and subsequently used to raise dykes, for example. The clean beet is then sliced in cutting machines. The sugar in these beet strips is removed in diffusion towers with warm water. The result is raw juice with a sugar concentration of 14 per cent (FAO, 2003). This is almost the same amount as in the beet itself. The extracted beet strips, now called pulp, are pressed or dried

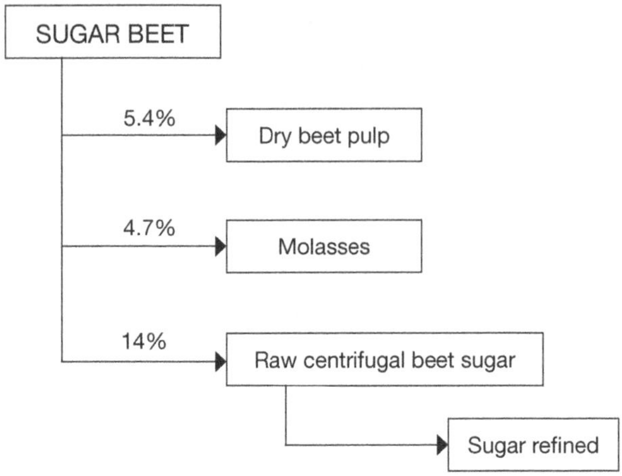

Figure III.1 Spanish refined sugar production (from sugar beet) diagram including product fractions

Source: Own elaboration based on FAO (2003)

and sold as animal fodder. The following stage in the production process is the purification of the raw juice. The raw juice is purified into thin juice with lime and carbon dioxide (CO_2). Lime and CO_2 are produced on the premises of the production plant in a lime oven from limestone and cokes. The lime absorbs all unwanted substances and this precipitates through the addition of CO_2. This solid matter is filtered off. It is a powerful, natural lime fertilizer that improves the structure of the soil and it is sold under the name Betacal SU. As the water evaporates the thin juice gradually becomes thick juice with a sugar content of approximately 70 per cent. Finally, so much water evaporates in the 'vacuum pans' that a saturated solution is obtained. Subsequently the crystallization process begins by adding fine sugar crystals that serve as seed material. By continuing to evaporate the water, these sugar crystals gradually develop to the required size. In centrifuges the crystal-clear sugar crystals are separated from the liquid (syrup) and, after drying, the sugar is stored in large silos. The syrup is called molasses and it serves as the raw material for the production of alcohol.

Sugar industry co-products are shown in the production diagram in Figure III.1. Beet pulp is dried and sold via the feed ingredient industry to dairy farmers who use dried pulp or store pressed pulp in silos and use the silage for milk and meat production. Pulp is also sold to farmers with sows, where it has a positive influence on environmental problems, as the dry matter content of manure produced by the sows is higher and the ammonia level in the pig house is lower. Experiments are also being done to feed sugar beet pulp to fattening pigs with promising results. The sugar industry molasses is sold to the alcohol industry and the co-product of this alcohol industry (vinasses) is used in the dairy feed industry, with a small part now used by farmers as a potassium fertilizer.

During the process described above, the use of water is limited as much as possible. The sugar factories use especially water from the beets. This is released in the production process as a condensate of evaporation. Sugar beets are more than 75 per cent water. Thus, during the sugar production, a surplus of water arises, originating from the sugar beets. After purification, this water becomes drained into surface water. During the washing of the beets, organic matter comes into the washing water and is purified. Besides aerobic purification, anaerobic purification also takes place in the methane engines, in which durable biogas is produced.

The water footprint of refined sugar has been estimated separately for the green, blue and grey components. This has been done in two steps: first for the intermediate raw centrifugal beet sugar and second for the refined sugar.

First, the blue water footprint of raw centrifugal beet sugar is estimated following the equation:

$$WF_{prod}[p] = \left(WF_{proc}[p] + \sum_{i=1}^{y} \frac{WF_{prod}[i]}{f_p[p,i]} \right) \times f_v[p] \quad \text{[volume/mass]} \qquad (68)$$

As described above, the process water footprint ($WF_{proc}[p]$) is equal to zero. The blue water footprint of the input sugar beet ($WF_{prod}[i]$) produced in Valladolid amounts to about 82m^3/ton (Appendix II). The product fraction ($f_p[p,i]$) in line with the sugar production diagram is 0.14ton/ton. And the value fraction ($f_v[p]$) amounts to about 0.89US$/US$ calculated as follows:

$$f_v[p] = \frac{price[p] \times w[p]}{\sum_{p=1}^{z} (price[p] \times w[p])} \quad [-] \qquad (69)$$

$$f_v[p] = \frac{price_{rawcentr.beetsugar} \times weight_{rawcentr.beetsugar}}{price_{drybeetpulp} \times w_{drybeetpulp} + price_{molasses} \times w_{molasses} + price_{rawcentr.beetsugar} \times w_{rawcentr.beetsugar}}$$

$$[-] \qquad (70)$$

All in all the blue water footprint of raw centrifugal beet sugar adds up to 524m^3/ton.

Second, the blue water footprint of refined sugar is calculated. Here again the process water footprint ($WF_{proc}[p]$) is equal to zero. The blue water footprint of the input raw centrifugal beet sugar ($WF_{prod}[i]$) is 524 m^3/ton. The product fraction ($f_p[p,i]$) in line with the sugar production diagram is 0.92 ton/ton and the value

Table III.1 *Green, blue and grey water footprint for sugar beet in Valladolid (Spain) (m³/ton)*

Process water footprint of sugar beet crop (m³/ton)				**Product water footprint of refined sugar (m³/ton)**			
$WF_{proc,green}$	$WF_{proc,blue}$	$WF_{proc,grey}$	WF_{total}	$WF_{proc,green}$	$WF_{proc,blue}$	$WF_{proc,grey}$	WF_{total}
15	82	22	120	107	570	152	829

fraction ($f_v[p]$) is 1US$/US$ since there is just one output product. Finally the blue water footprint of refined sugar produced in Valladolid is 570m³/ton. The green and grey water footprint are calculated in a similar way (Table III.1).

Appendix IV

Examples of Grey Water Footprint Calculations

Example 1: Grey water footprint from point source pollution

Consider a water-using process as depicted below. The abstraction is 0.10 m^3/s; the effluent is 0.09 m^3/s, a bit smaller than the abstraction because some of the water that was abstracted evaporates during the process, so that it is not returned to the freshwater body. The natural concentration of a certain chemical in the freshwater body (c_{nat}) is 0.5 g/m^3, but the actual concentration (c_{act}) at the point of abstraction is already 1 g/m^3, due to polluting activities upstream. The concentration of the chemical in the effluent (c_{effl}) is 15 g/m^3. The maximum acceptable concentration in the water body (c_{max}) is 10 g/m^3. The (additional) load of this process to the freshwater body is equal to: $0.09 \times 15 - 0.1 \times 1 = 1.25$ g/s. The associated grey water footprint is: $1.25 / (10 - 0.5) = 0.13$ m^3/s.

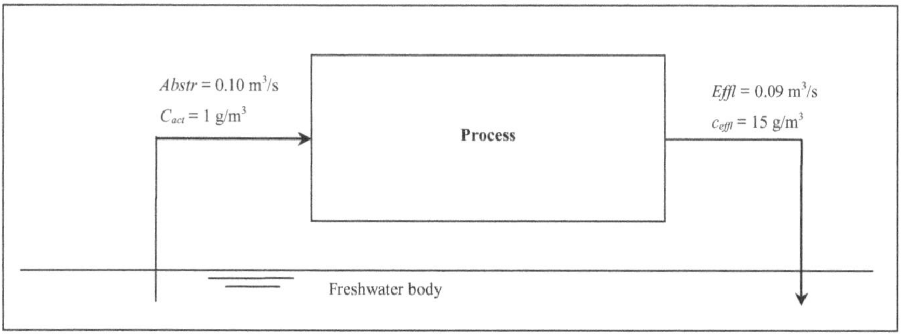

In this example, the concentration in the effluent is larger than the maximum acceptable concentration in the freshwater body. A smart process manager, who however doesn't really understand environmental impacts, decides to abstract more water in order to be able to dilute the effluent such that the concentration in the effluent becomes equal to the maximum acceptable concentration. The volume of abstraction is increased from 0.10 to 0.15 m^3/s. The effluent volume becomes 0.14 m^3/s, because the evaporation loss in the process remains the same. The concentration of the chemical in the effluent is now diluted to a level of 10 g/m^3. The (additional) load of the freshwater body remains the same as before: 0.14 × 10 – 0.15 × 1 = 1.25 g/s. The associated grey water footprint thus remains the same as well: 1.25 / (10 – 0.5) = 0.13 m^3/s. It appears that what the process manager has done looks nice – because the concentration in the effluent has been reduced to 'acceptable level' – but that for the receiving freshwater it makes no difference – because the load and grey water footprint remain the same.

Finally, it is decided to stop the intake of additional water again, so that abstraction is 0.10 m^3/s again. Instead, one treats the wastewater before it is disposed into the environment. During the treatment, a large fraction of the chemicals is removed from the effluent. The treatment is designed such that there is no evaporation loss during the treatment, so that the effluent volume remains 0.09 m^3/s. However, the concentration of the chemical in the effluent (c_{effl}) is reduced from 15 g/m^3 down to 2 g/m^3. The (additional) load of the process to the freshwater body is now equal to: 0.09 × 2 – 0.1 × 1 = 0.08 g/s. The associated grey water footprint is: 0.08 / (10 – 0.5) = 0.0084 m^3/s. Although the concentration of the chemical in the effluent is below the maximum allowable concentration in the freshwater body, one can see that the grey water footprint is not zero. The reason is that the concentration in the effluent is still beyond the natural concentration in the water body, so that the process still puts some claim on the waste assimilation capacity of the freshwater body.

Example 2: Calculating the water pollution level at different scales

Consider a catchment that can be subdivided into three sub-catchments as schematized in the picture below. There are two upstream sub-catchments that drain into a third, downstream sub-catchment. In each sub-catchment there is a total load of a certain chemical of 2000 kg in a certain month. With a natural concentration of zero and a maximum acceptable concentration for this chemical of 0.01 kg/m^3, the grey water footprint in each sub-catchment can be calculated as 2000 / (0.01 – 0) = 0.2 million m^3 in the month under consideration. In that month, runoff from sub-catchment 1 amounts to 1 million m^3, runoff from sub-catchment 2 is 0.2 million m^3 and runoff from sub-catchment 3 is 0.8 million m^3. Presuming that the residence time of the water in the catchment is low, the total runoff from the catchment in the month considered is equal to the sum of the runoff volumes from the three sub-catchments, i.e. 2 million m^3. One can calculate a 'water pollution level' per sub-catchment for the month considered

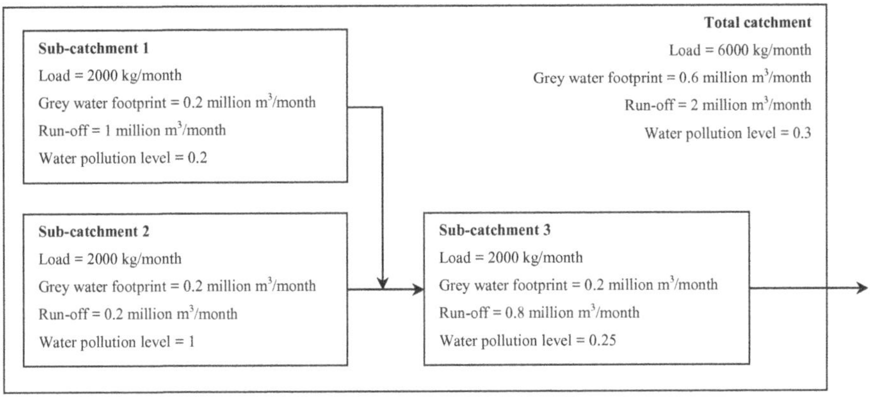

by taking the ratio of the grey water footprint to the run-off. The results are shown in the picture below. In sub-catchment 2 the water pollution level is 1, which means that the waste assimilation capacity in this sub-catchment has been fully consumed. This is not the case in the other two sub-catchments. When looking at the catchment as a whole, one arrives at a water pollution level of 0.3. This illustrates the fact that, when pollution is unequally distributed within a catchment, the hotspots become visible only when the resolution level of the analysis is sufficiently high.

This example can also be used to explain why grey water footprints are measured based on the (human-induced) loads that enter into freshwater bodies, and not on the basis of the loads that can finally be measured downstream in the river at the point where it leaves the catchment. Suppose that the load added in sub-catchment 1, on the way through sub-catchments 1 and 3, is partly broken down under influence of biochemical processes in the river, so that only 80 per cent of the original load finally leaves sub-catchment 3. Assume that the same holds for the load that was disposed into the water in sub-catchment 2. Finally, assume that 90 per cent of the load brought into sub-catchment 3 finally leaves that sub-catchment. Then we can calculate that the chemical load in the river at the most downstream point amounts to 5000 kg. This is less than the 6000 kg that was originally added to the waters of the catchment, distributed over the catchment. Calculating the grey water footprint or the water pollution level on the basis of downstream loads instead of loads as they enter the system gives a false impression of the situation. This becomes most clear if we slightly change the above example. Assume now that the loads in sub-catchments 1-3 amount to 10000, 2000 and 8000 kg/month, respectively. This means that we calculate a water pollution level of 1 for each sub-catchment. If that is the case for each sub-catchment, then this must also be the case for the catchment as a whole. When, however, one would look at the chemical load at the outlet of sub-catchment 3, then one finds a load of 0.8×(10000+2000) + 0.9×8000 = 16800 kg/month. If we assess the water pollution level in the catchment as a whole on the basis of the load in the outlet, one calculates a misleading overall pollution level of 0.84.

Appendix V

Environmental Flow Requirements

In the framework of water footprint discussions it is crucial to have standards on environmental flow requirements. According to the Brisbane Declaration, which was drafted at the 10th International Riversymposium and Environmental Flows Conference, held in Brisbane, Australia, in 2007, environmental flow requirements are defined as the 'quantity, timing and quality of water flows required to sustain freshwater and estuarine ecosystems and the human livelihood and well-being that depend on these ecosystems' (Poff et al, 2010).

When we are interested in the environmental impacts of the blue water footprint (consumptive water use of run-off), it is crucial to know the environmental flow requirements in the catchment where the blue water footprint is located. We will focus in this context on the environmentally required quantity and timing of water flows. Natural run-off from a catchment (R_{nat}) minus environmental flow requirements (*EFR*) is what is available for human use. Blue water availability (WA_{blue}) is thus defined as:

$$WA_{blue} = R_{nat} - EFR \quad \text{[volume/time]} \tag{71}$$

The blue water footprint (WF_{blue}) in a catchment needs to be compared to WA_{blue}. When WF_{blue} approaches or exceeds WA_{blue} there is a reason for concern. The natural run-off can be estimated as the actual run-off plus the blue water footprint. We know run-off for many catchments in the world and if no empirical data are available we have model estimates. The time resolution is sometimes daily, but at least we generally know run-off on a monthly basis. Water footprint data have so far mostly been presented on an annual basis, but behind those estimates is always information about its course over time, because water footprint calculations are based on irrigation water use calculations with a time step of 1 to 10 days. Comparing WF_{blue} to WA_{blue} can be done on an annual basis, but that is very crude and insensitive to what actually happens throughout the year, so it is better to make this comparison on, for example, a monthly basis. There is sufficient literature to conclude that establishing *EFR* in a particular catchment will always be an elaborate job. It is tempting to have a simple, generic, easily applicable standard on estimating *EFR*, so that one can easily assess the environmental impact of a blue water footprint in an arbitrary catchment in the world. The broad literature on environmental flow requirements provides many useful methods, guidelines and examples, but there is only one worldwide study on environmental flow requirements based on a simple rule and readily available data: the study by Smakhtin et al (2004). The good thing about this study is that it offers what many practitioners want (easy method, clear numbers, world-coverage); the Smakhtin-map frequently features in business reports and presentations. The downside is that the method gives annual instead of monthly values for *EFR* and that many experts do not agree with the calculation rules, parameters used and the resultant *EFR* estimates. According to Arthington et al (2006), the Smakhtin-method greatly underestimates environmental flow requirements.

For practical purposes it is proposed here to work towards a simple (based on readily available data) and generic (worldwide applicable) method to establish environmental flow requirements for catchments with a low temporal resolution but high enough to capture the main variations within a year. The estimates obtained with this method can function as default *EFR* estimates in cases where more advanced estimates are not yet available. It should be stressed that the simple generic method would give first estimates only, to be replaced by better estimates when possible. For that purpose one could rely, for example, on the ELOHA framework for establishing environmental flow requirements: an advanced framework being proposed by some of world's top experts in this field (Poff et al, 2010). This method is money- and labour-intensive, so it will take at least several years before we will have a worldwide coverage of *EFR* estimates based on this approach.

For the time being, the following simple generic rule for establishing environmental flow requirements is proposed:

1. For each month of the year, the mean monthly run-off in developed condition is in a range ±20 per cent of the mean monthly flow as would happen under undeveloped condition, and
2. For each month of the year, the mean monthly base flow is in a range ±20 per cent of the mean monthly base flow as would happen under undeveloped condition.

Mean monthly flows are often available through river flow measurements, and if not we can use model estimates. The term 'base flow' refers to the groundwater contribution to a river, which can be estimated based on for instance a 10-year river flow record.

In order to create more detail it is proposed to distinguish 'levels of river basin modification'. Referring to the deviation (Δ) of the mean monthly flows under developed conditions to those under undeveloped conditions, the following scheme can be used:

$\Delta < \pm 20\%$	unmodified or slightly modified	river status A
$\pm 20\% < \Delta < \pm 30\%$	moderately modified	river status B
$\pm 20\% < \Delta < \pm 40\%$	significantly modified	river status C
$\Delta > \pm 40\%$	seriously modified	river status D

Today, how many basins will fall in the four categories A to D? The majority of non-dam-regulated rivers will fall in status A. The dam-regulated rivers will fall in categories B-D. The 20 per cent rule is regarded as a 'precautionary default *EFR*'. The above boundaries can be called 'thresholds for potential concern'. This terminology better reflects the fact that the above boundaries are indicative rather than decisive.

The appropriate spatial scale for establishing *EFR* is the catchment level. *EFR* at river basin level can be derived as the sum of *EFR* values of the catchments that together constitute the river basin. Given that *EFR* can best be expressed at catchment level, ideally the water footprint is specified at that detail as well. Ideally, water footprint accounting is done in a spatial explicit way using a geographic information system (GIS), so in that case one can always localize the water footprint rather precise.

The local impact of a blue water footprint in a river can be quantified by counting the (average) number of months in a year that the environmental flow requirements in the river are not met and by considering the degree to which the environmental flow requirements are violated. It is not said that the blue water footprint of the activity considered is fully responsible for the violation of the environmental flow requirements, because it is the sum of the blue water footprints of all activities that results in violation. Therefore one can also look at the relative contribution of the activity considered.

The above simple method is based on initial thoughts from some water resource experts (personal communication between Jay O'Keeffe, UNESCO-IHE; Brian Richter, TNC; Stuart Orr, WWF; Arjen Hoekstra, University of Twente). We need agreement among and support from the broad community of *EFR* experts on this simple generic method, because undoubtedly the method will be criticized, which is understandable given both the diverse interests (environment versus water users) and the scientific difficulty to translate the actual complexity towards simple rules. However, possible criticism has not withhold experts from setting simple toxicity and water quality standards, so why would it withhold us from establishing *EFR* standards? Quantifying environmental flow requirements is essential to be able to let them count in assessing the impacts of blue water consumption.

Appendix VI

Frequently Asked Questions

Practical questions

1. Why should we bother about our water footprint?

Freshwater is a scarce resource; its annual availability is limited and demand is growing. The water footprint of humanity has exceeded sustainable levels in several places and is unequally distributed among people. Good information regarding water footprints of communities and businesses will help us to understand how we can achieve a more sustainable and equitable use of freshwater. There are many spots in the world where serious water depletion or pollution takes place: rivers running dry, dropping lake and groundwater levels and endangered species resulting from contaminated water. The water footprint helps to show the link that exists between our daily consumption of goods and the problems of water depletion and pollution that exist elsewhere, in the regions where our goods are produced. Nearly every product has a smaller or larger water footprint, which is of interest for both the consumers that buy those products and the businesses that produce, process, trade or sell those products in some stage of their supply chain.

2. Why should my business bother about its water footprint?

First of all, environmental awareness and strategy is often part of what a business regards as its 'corporate social responsibility'. Reducing the water footprint can be part of the environmental strategy of a business, just like reducing the carbon footprint. Second, many businesses actually face serious risks related to freshwater shortage in their operations or supply chain. What is a brewery without a secure water supply or how can a company that makes jeans survive without continued supply of water to the cotton fields? A third reason to do water footprint accounting and formulate measures to reduce the corporate water footprint is to anticipate regulatory control by governments. In the current stage, it is not so clear how governments will respond, but obviously regulations in some sectors of business may be expected. Finally, some businesses see a corporate water footprint strategy also as an instrument to reinforce the corporate image or to strengthen the brand name.

3. What can consumers do to reduce their water footprint?

Consumers can reduce their *direct* water footprint (home water use) by installing water-saving toilets, applying a water-saving showerhead, turning off the tap during teeth-brushing, using less water in the garden and by not disposing of medicines, paints or other pollutants down the sink. The *indirect* water footprint of a consumer is generally much larger than the direct one. A consumer has basically two options to reduce his/her indirect water footprint. One option is to substitute a consumer product that has a large water footprint by a different type of product that has a smaller water footprint. Examples include eating less meat or becoming vegetarian, drinking tea instead of coffee or, even better, drinking plain water. Not wearing cotton but clothes made from artificial fibres would save a lot of water. But this approach has limitations, because many people do not easily shift from eating meat to being vegetarian and people like their coffee and cotton. A second option is to stick to the same consumption pattern but to select the cotton, beef or coffee that has a relatively low water footprint or that has its footprint in an area that does not have high water scarcity. This requires, however, that consumers have the proper information to make that choice. Since this information is generally not available in the world of today, an important thing consumers can do now is ask for product transparency from businesses and regulation from governments. When information is available on the impacts of a certain article on the water system, consumers can make conscious choices about what they buy.

4. What can businesses do to reduce their water footprint?

Businesses can reduce their *operational* water footprint by saving water in their own operations and bringing water pollution to zero. Keywords are: avoid, reduce, recycle and treat before disposal. For most businesses, however, the *supply chain* water footprint is much larger than the operational footprint. It is therefore crucial that businesses address that as well. Achieving improvements in the supply chain may be more difficult – because not under direct control – but they may be more effective. Businesses can reduce their supply-chain water footprint by making supply agreements with certain standards with their suppliers or by simply changing to another supplier. In many cases it probably means quite something, because the whole business model may need to be transformed in order to incorporate or better control supply chains and to make supply chains fully transparent to consumers. Among the various alternative or supplementary tools that can help improving transparency are: setting quantitative water-footprint reduction targets, benchmarking, product labelling, certification and water footprint reporting.

5. Why should governments make national water footprint accounts?

Traditionally, countries formulate national water plans by looking at how to satisfy water users. Even though countries nowadays consider options to reduce water demand in addition to options to increase supply, they generally do not include the global dimension of water management. In this way they do not explicitly consider options to save water through import of water-intensive products. In addition, by looking only at water use in their own country, most governments have a blind spot to the issue of sustainability of national consumption. As a matter of fact, many countries have significantly externalized their water footprint without looking at whether the imported products are related to water depletion or pollution in the producing countries. Governments can and should engage with consumers and businesses to work towards sustainable consumer products. National water footprint accounting should be a standard component in national water statistics and provide a basis to formulate a national water plan and river basin plans that are coherent with national policies with respect to the environment, agriculture, industry, energy, trade, foreign affairs and international cooperation.

6. When is my water footprint sustainable?

As a consumer, your water footprint is sustainable when (i) the total remains below your fair share in the world, (ii) no component of your total water footprint is located in a hotspot, and (iii) no component of your total water footprint can be reduced or avoided altogether at reasonable societal cost.

7. How can I offset my water footprint?

This is a question often posed by people that are familiar with the idea of carbon offsetting. In the case of carbon it does not matter where mitigating measures are taken, so one can offset own CO_2 emissions by helping to reduce CO_2 emissions or enhancing carbon sequestration elsewhere. In the case of water, this is different, because water depletion or pollution in one place cannot be compensated by whatever measure in another place. The focus should therefore be on the reduction of one's own water footprint, most urgent at the places where and times when this water footprint causes problems. We should do all that is 'reasonably possible' to reduce our own water footprint, both the direct and indirect one. This holds for both consumers and businesses. Only in the second instance, when everything has been done to reduce the water footprint, can one consider offsetting. This means that the residual water footprint is offset by making a 'reasonable investment' in establishing or supporting projects that aim at a sustainable, equitable and efficient use of water in the catchment where the residual water footprint is located. The terms 'reasonably possible' and 'reasonable investment' include normative elements that need further quantitative specification and about which we need to reach societal consensus.

8. I already pay for the water, is that not enough?

Generally the price paid for blue water is far below its real economic cost. Most governments subsidize blue water supply on a huge scale by investing in infrastructure like dams, canals, distribution systems and wastewater treatment. These costs are often not charged to the water users. As a result, there is insufficient economic incentive for water users to save water. Besides, due to the public character of water, water scarcity is generally not translated into an additional component in the price of goods and services that are produced with the water, as happens naturally in the case of private goods. Finally, water users generally do not pay for the negative impacts that they cause on downstream people or ecosystems.

9. Why should we reduce green water footprints?

One can argue that the rain comes for free; if humans do not use green water to produce food, fibres, timber or bio-energy, it will evaporate anyway. There are, however, two good reasons for reducing green water footprints. The first is: rain is free, but not unlimited. In fact, green water is a scarce resource, just like blue water, particularly in some places and during certain parts of the year. Since part of the land in any river basin has to be reserved for nature, automatically a certain amount of green water is not available for agriculture. In catchments where green water is scarce, increasing the productivity of green water (in other words, reducing the green water footprint of a product) is crucial in order to gain optimum production given the green water constraint. The second reason is that increased production based on green water resources reduces the need for production with blue water resources. This is why reducing green water footprints is also useful in areas where green water is abundantly available. Better use of rain in areas where rain is sufficient will increase the worldwide production of rain-based products, which reduces the need to produce irrigation-based products in water-scarce areas.

10. Why should we reduce blue water footprints in areas with sufficient run-off?

At first sight, it seems necessary to reduce blue water footprints only in catchments where blue water availability is insufficient. Focussing on water-scarce areas, however, is inadequate. Inefficient water use in water-abundant places implies that production per unit of water can be increased, which is important because increased production of water-intensive goods in water-abundant places means that production of those goods in water-scarce places can be reduced. Lowering the water footprint per unit of product in water-abundant areas thus contributes to the possibility to reduce the overall water footprint in water-scarce areas. Another reason for reducing blue water footprints in water-abundant areas is that the allocation of blue water to one purpose withdraws from the possibility to allocate it for another purpose. The water footprints of water-intensive and luxury products such as meat, bio-energy or cut flowers can press in catchments where water is abundantly available and where local environmental flow requirements are fulfilled, but the global implications of these water footprints are that less water remains to be allocated to other purposes, such as growing cereal crops to fulfil basic food demand. Reducing the blue water footprint of a specific product in a water-abundant area thus creates the possibility to produce more of that specific product or to allocate the saved water to another product.

11. What are reasonable water footprint reduction targets?

There is no general answer to this question, because it depends on the product, available technology, local context and so on. Besides, one has to keep in mind that the question includes a normative element, which implies that it needs to be answered in a societal-political context. A few general things can be said, however. First of all, one has to distinguish between reduction targets with respect to the green, blue and grey water footprint. As for the grey water footprint, which refers to water pollution, one can demand a reduction to zero for all products, at least in the long term. Pollution is not necessary. A zero grey water footprint can be achieved by prevention, recycling and treatment. Only thermal pollution (by water use for cooling) is difficult to reduce to zero, but even this sort of pollution can be (largely) prevented by recapturing the heat. The blue water footprint in the agricultural stage of products can substantially be brought down by reducing consumptive water losses, increasing blue water productivity and relying more on rain-fed agriculture. In the industrial stage it will depend very much on the sector and what has already been done. Technologically, industries can fully recycle water, so that the blue water footprint everywhere can be reduced to the amount of water that is actually being incorporated into the product. Benchmarks can be developed for specific products by taking the performance of the best producers as a reference. Green water footprints in agriculture can often be reduced very substantially by using the green water resources more efficiently, in other words, by increasing the green water productivity. Increased production based on green water resources in one place will reduce the need for production with blue water resources elsewhere. A general rule for any water footprint mitigation strategy is to avoid the water footprint pressing in areas or times where environmental water needs are violated. A final rationale for a water footprint mitigation strategy can be the fair sharing of water resources. This may be the basis for water footprint reduction particularly for large water users.

12. Is the water footprint similar to the carbon footprint?

The two concepts nicely complement each other, each concept addressing another environmental issue: the carbon footprint addresses the issue of climate change, the water footprint relates to the issue of freshwater scarcity. In both cases, a supply chain perspective is promoted. There are also differences, however. For a carbon emission it does not matter where it happens, but for a water footprint is does matter. A carbon emission in one place can be offset by carbon emission reduction or sequestration in another place, which is not true for water: one cannot reduce the local impact of water use in one place by saving water in another place.

13. Freshwater can be obtained by desalinating seawater, so why is water scarce?

Desalination of salt or brackish water can only be a solution for freshwater scarcity in a limited number of applications, not because one cannot obtain the right quality of water for all purposes, but because desalination requires energy, another scarce resource. In fact, desalination is a way of substituting one scarce resource (freshwater) for another (energy). If, at a certain spot, the freshwater issue is pressing even more than the energy issue, one can decide in favour of desalination, but in general it does not make sense to propose desalination as a general solution to freshwater scarcity. Besides, apart from the energy argument, desalination is still expensive, too expensive for use in agriculture where most of the water is used. Finally, salt or brackish water is only available along coasts, which means that bringing desalinated water elsewhere would imply additional costs (again including energy).

14. Should products get a water label?

In a world where many products are related to water depletion and pollution it is very useful to make the history of products more transparent. It is good to have the facts publicly available, so the consumer has a choice. Information can be provided on a label or can be made available through the internet. This is most useful for products that are often associated with large effects on water, such as products that contain cotton or sugar. For consumers it would be helpful to integrate a water label in broader labels that include other issues as well, such as energy and fair trade. Ideal would be a world in which we do not need labels because we can trust that all products meet strict criteria. If a water label is considered, the question is what should be on the label. One could put the total water footprint of the product on a label, which is functional only for raising awareness among consumers, not for enabling the consumer to make a well-informed choice between two products. For supporting good product choice, one would also need to specify the green-blue-grey components and mention the degree in which the product's water footprint relates to violation of local environmental flow requirements or ambient water quality standards. For example, three quarters of the water footprint is situated in areas where environmental flow requirements or ambient water quality standards are met, but the other quarter of the total water footprint is in areas where those norms are violated. Whether a product is 'good' or not from a water resources point of view depends on a whole range of criteria, including whether plans are in place to achieve continued improvements along the production chain. In the end, labelling of products is a partial solution at best. As a means of awareness-raising and basis for product choice, it can be functional, but it is just one way of providing product transparency, restricted by the practical problem that a label can contain limited information only. Besides, real water footprint reduction will not occur just by providing information on a label.

Technical questions

1. What is a water footprint?

The water footprint of a product is an empirical indicator of how much water is consumed and polluted, when and where, measured over the whole supply chain of the product. The water footprint is a multidimensional indicator, showing volumes but also making explicit the type of water use (consumptive use of rainwater, surface water or groundwater, or pollution of water) and the location and timing of water use. The water footprint of an individual, community or business, is defined as the total volume of freshwater that is used to produce the goods and services consumed by the individual or community or produced by the business. The water footprint shows human appropriation of the world's limited freshwater resources and thus provides a basis for discussing water allocation and issues that relate to sustainable, equitable and efficient water use. Besides, the water footprint forms a basis for assessing the impacts of goods and services at catchment level and formulating strategies to reduce those impacts.

2. What is new about the water footprint?

Traditionally, statistics on water use focus on measuring 'water withdrawals' and 'direct water use'. The water footprint accounting method takes a much broader perspective. First of all, the water footprint measures both direct and indirect water use, where the latter refers to the water use in the supply chain of a product. The water footprint thus links final consumers and intermediate businesses and traders to the water use along the whole production chain of a product. This is relevant, because generally the direct water use of a consumer is small if compared to its indirect water use and the operational water use of a business is generally small if compared to the supply chain water use. So the picture of the actual water dependency of a consumer and business can change radically. The water footprint method differs further in that it looks at water consumption (as opposed to withdrawal), where consumption refers to the part of the water withdrawal that evaporates or gets incorporated into a product. Besides, the water footprint goes beyond looking at blue water use only (use of ground and surface water). It also includes a green water footprint component (use of rainwater) and a grey water footprint component (polluted water).

3. Is the water footprint more than a nice metaphor?

The term 'footprint' is often used as a metaphor to refer to the fact that humanity appropriates a significant proportion of the available natural resources (land, energy, water). However, just like the 'ecological footprint' and the 'carbon footprint', the 'water footprint' is more than a metaphor: there is a rigorous accounting framework with well-defined measurable variables and well-established accounting procedures to calculate the water footprints of products, individual consumers, communities, nations or businesses. We discourage people to use the water-footprint concept as a metaphor, because its strength lies in its effectiveness when used in a context of strict accounting and measurable reduction targets.

4. Water is a renewable resource, it remains in the cycle, so what is the problem?

Water is a renewable resource, but that does not mean that its availability is unlimited. In a certain period, precipitation is always limited to a certain amount. The same holds to the amount of water that recharges groundwater reserves or flows through a river. Rainwater can be used in agricultural production and water in rivers and aquifers can be used for irrigation or industrial or domestic purposes. But one cannot use more water than is available. One cannot take more from a river than its flow in a certain period and, in the long term, one cannot take more water from lakes and groundwater reservoirs than the rate with which they are recharged. The water footprint measures the amount of water available in a certain period that is consumed (evaporated) or polluted. In this way, it provides a measure of the amount of available water appropriated by humans. The remainder is left for nature. The rainwater not used for agricultural production is left to sustain natural vegetation. The groundwater and surface water flows not evaporated for human purposes or polluted are left to sustain healthy aquatic ecosystems.

5. Is there agreement regarding how to measure a water footprint?

The methods for water footprint accounting have been published in peer-reviewed scientific journals. In addition, there are also practical examples available of how one can apply the methods to calculate the water footprint of a specific product, an individual consumer, a community or a business or organization. In a generic sense, there is agreement regarding the definition and calculation of a water footprint. However, every time one applies the concept in a situation not encountered before, new practical questions arise. These are questions such as: what should be included and what can be excluded, how to deal with situations where the supply chain cannot be properly traced, what water quality standards to use when calculating the grey water footprint and so on. Discussion therefore focuses on how to handle those practical issues.

6. Why distinguish between a green, blue and grey water footprint?

Freshwater availability on Earth is determined by annual precipitation above land. One part of the precipitation evaporates and the other part runs off to the ocean through aquifers and rivers. Both the evaporative flow and the run-off flow can be made productive for human purposes. The evaporative flow can be used for crop growth or left for maintaining natural ecosystems; the green water footprint measures what part of the total evaporative flow is actually appropriated for human purposes. The run-off flow – the water flowing in aquifers and rivers – can be used for all sorts of purposes, including irrigation, washing, processing and cooling. The blue water footprint measures the volume of groundwater and surface water consumed (in other words, withdrawn and then evaporated or incorporated into a product). The grey water footprint measures the volume of water flow in aquifers and rivers polluted by humans. In this way, the green, blue and grey water footprint measure different sorts of water appropriation. When necessary, one can further classify the water footprint into more specific components. In the case of the blue water footprint, it can be considered relevant to distinguish between surface water, renewable groundwater and fossil groundwater. In the case of the grey water footprint, it can be considered valuable to distinguish between different sorts of pollution. In fact, preferably, these more specific pieces of information are always underlying the aggregate water footprint figures.

7. Why should we look at the total green water footprint of a crop? Why not look at the additional evaporation if compared to evaporation from natural vegetation?

It depends on what question one would like to address. The green water footprint measures total evaporation and is meant to feed the debate regarding the allocation of water to different purposes in a context of limited availability. Information about increased or reduced evaporation is relevant from the perspective of catchment hydrology and potential downstream effects. Research has shown that crops can sometimes result in increased evaporation when compared to natural vegetation (particularly in the period of rapid crop growth), and other times in reduced evaporation (for example, because of soil deterioration or reduced aboveground biomass). In many cases the differences are not very significant at basin scale. The change in evaporation is interesting from the perspective of catchment hydrology and potential downstream effects, but not for the debate on how limited freshwater resources are allocated over different purposes. The water footprint is designed for the latter debate. The purpose of the green water footprint is to measure human's appropriation of the evaporative flow, just like the blue/grey water footprint aims to measure human's appropriation of the run-off flow. The green water footprint measures the part of the evaporated rainwater that has been appropriated by human beings and is therefore not available for nature. The water footprint thus expresses the cost of a crop in terms of its total water use.

8. Is it not too simplistic to add all cubic metres of water used into one aggregate indicator?

The aggregate water footprint of a product, consumer or producer shows the total volume of fresh water appropriated (consumed or polluted). It serves as a rough indicator, instrumental in raising awareness and for getting an idea of where most of the water goes. The water footprint can be presented as one aggregate number, but in fact it is a multidimensional indicator of water use, showing different sorts of water consumption and pollution as a function of space and time. For developing strategies for sustainable water use, one will need to use the more detailed layer of information embedded in the composite water footprint indicator.

9. Should we not weigh the different water footprint components based on their local impact?

The idea of 'weighing factors' sounds like an attractive idea, because not every cubic metre of water used has the same local impact. However, we strongly discourage this approach for three reasons. First, weighing is and will always remain very subjective, because there are many different sorts of (environmental, social and economic) impacts, some of which cannot even be easily quantified. Second, impacts are always fully local-context dependent, which means it is impossible to design universally valid weighing factors. The impact of one cubic metre of water withdrawn from one particular point in a river at a certain point in time depends on the characteristics of that river, such as the volume and variability of water flow in the river, the competition over water at that point in the river at that particular moment and the effects of withdrawal on downstream ecosystems and other users. Third, and most important, the volumetric water footprint figures actually contain crucial information, which is obscured when weighed. Water footprints refer to actual volumes of water appropriated, which is crucial information in itself, because in a world in which freshwater resources are scarce, it is important to know what volumes are allocated to different purposes. That local impacts of water consumption or pollution differ is another issue. In order to properly address the fact that different water footprint components do indeed have different local impacts, we emphasize that the water footprint is a multidimensional indicator, showing volumes, but also the type of water use and the locations and timing of water use. 'Water footprint accounting' means that one quantifies the water footprint in all its details. This forms the proper basis for a local impact assessment, in which one assesses the various impacts for each separate water footprint component in time and space. Obviously, the local impact assessment will show that the impact is different for each separate water footprint component. For formulating water policy aimed to reduce water footprint impacts, it is more useful to know how different water footprint components link to various impacts than to have a weighed water footprint impact index. The risk of making a seemingly advanced weighed water footprint impact index is that such an index hides all information related to impacts instead of making the impacts explicit.

10. How does water footprint accounting relate to life cycle assessment?

The water footprint of a product can be an indicator in the life cycle assessment (LCA) of a product. Being applied in an LCA is one of the many applications of the water footprint. In a global context, the water footprint is a relevant indicator of the how much of the globe's scarce freshwater resources are used for a certain product. In a more local context, the spatiotemporally explicit water footprint can be overlaid with a water-stress map in order to arrive at a spatiotemporally explicit water footprint impact map. The various impacts should subsequently be weighed and aggregated in order to arrive at an aggregated water footprint impact index. For LCA, an important question is how different sorts of natural resource use and environmental impacts can be aggregated – which is a specific requirement for LCA and not relevant to other applications of the water footprint. Other applications of the water footprint are, for example, identifying hotspot areas of the water footprints of certain products, consumer groups or businesses, and formulating response strategies to reduce water footprints and mitigate associated impacts. For these purposes, aggregation is not functional, because specification in type of water and space-time is essential in those applications.

11. How does the water footprint relate to ecological and carbon footprints?

The water-footprint concept is part of a larger family of concepts that have been developed in the environmental sciences over the past decade. A 'footprint' in general has become known as a quantitative measure showing the appropriation of natural resources or pressure on the environment by human beings. The ecological footprint is a measure of the use of bioproductive space (hectares). The carbon footprint measures the amount of greenhouse gases (GHGs) produced, measured carbon dioxide equivalents (in tonnes). The water footprint measures water use (in cubic metres per year). The three indicators are complementary, since they measure completely different things. Methodologically, there are many similarities between the different footprints, but each has its own peculiarities related to the uniqueness of the substance considered. Most typical for the water footprint is the importance of specifying space and time. This is necessary because the availability of water highly varies in space and time, so that water appropriation should always be considered in its local context.

12. What is the difference between water footprint and virtual water?

The water footprint is a term that refers to the water used to make a product. In this context we can also speak about the 'virtual water content' of a product instead of its 'water footprint'. The water footprint concept, however, has a wider application. We can, for example, speak about the water footprint of a consumer by looking at the water footprints of the goods and services consumed, or about the water footprint of a producer (business, manufacturer, service provider) by looking at the water footprint of the goods and services produced. Furthermore, the water footprint concept does not simply refer only to a water volume, as in the case of the term 'virtual water content' of a product. The water footprint is a multidimensional indicator, not only referring to a water volume used, but also making explicit where the water footprint is located, what source of water is used and when the water is used. The additional information is crucial in order to assess the local impacts of the water footprint of a product.

References

Acreman, M. and Dunbar, M. J. (2004) 'Defining environmental river flow requirements: A review', *Hydrology and Earth System Sciences*, vol 8, no 5, pp861–876

Alcamo, J. and Henrichs, T. (2002) 'Critical regions: A model-based estimation of world water resources sensitive to global changes', *Aquatic Sciences*, vol 64, no 4, pp352–362

Aldaya, M. M. and Hoekstra, A. Y. (2010) 'The water needed for Italians to eat pasta and pizza', *Agricultural Systems*, vol 103, pp351–360

Aldaya, M. M. and Llamas, M. R. (2008) 'Water footprint analysis for the Guadiana river basin', Value of Water Research Report Series No 35, UNESCO-IHE, Delft, Netherlands, www.waterfootprint.org/Reports/Report35-WaterFootprint-Guadiana.pdf

Aldaya, M. M., Allan, J. A. and Hoekstra, A. Y. (2010a) 'Strategic importance of green water in international crop trade', *Ecological Economics*, vol 69, no 4, pp887–894

Aldaya, M. M., Garrido, A., Llamas, M. R., Varelo-Ortega, C., Novo, P. and Casado, R. R. (2010b) 'Water footprint and virtual water trade in Spain', in A. Garrido and M. R. Llamas (eds) *Water Policy in Spain*, CRC Press, Leiden, Netherlands, pp49–59

Aldaya, M. M., Muñoz, G. and Hoekstra, A. Y. (2010c) 'Water footprint of cotton, wheat and rice production in Central Asia', Value of Water Research Report Series No 41, UNESCO-IHE, Delft, Netherlands, www.waterfootprint.org/Reports/Report41-WaterFootprintCentralAsia.pdf

Aldaya, M. M., Martinez-Santos, P. and Llamas, M. R. (2010d) 'Incorporating the water footprint and virtual water into policy: Reflections from the Mancha Occidental Region, Spain', *Water Resources Management*, vol 24, no 5, pp941–958

Allan, J. A. (2003) 'Virtual water – the water, food, and trade nexus: Useful concept or misleading metaphor?', *Water International*, vol 28, no 1, pp106–113

Allen, R. G., Pereira, L. S., Raes, D. and Smith, M. (1998) 'Crop evapotranspiration: Guidelines for computing crop water requirements', FAO Irrigation and Drainage Paper 56, Food and Agriculture Organization, Rome

ANZECC and ARMCANZ (Australian and New Zealand Environment and Conservation Council and Agriculture and Resource Management Council of Australia and New Zealand) (2000) 'Australian and New Zealand guidelines for fresh and marine water quality', ANZECC and ARMCANZ, www.mincos.gov.au/publications/australian_and_new_zealand_guidelines_for_fresh_and_marine_water_quality

Arthington, A. H., Bunn, S. E., Poff, N. L. and Naiman, R. J. (2006) 'The challenge of providing environmental flow rules to sustain river ecosystems', *Ecological Applications*, vol 16, no 4, pp1311–1318

Austrian Federal Ministry of Agriculture, Forestry, Environment and Water Management (2010) 'BGBl 2010 II Nr. 99: Verordnung des Bundesministers für Land- und Forstwirtschaft, Umwelt und Wasserwirtschaft über die Festlegung des ökologischen Zustandes für Oberflächengewässer (Qualitätszielverordnung Ökologie Oberflächengewässer – QZV Ökologie OG)'

Barton, B. (2010) 'Murky waters? Corporate reporting on water risk, A benchmarking study of 100 companies', Ceres, Boston, MA, www.ceres.org/Document.Doc?id=547

Batjes, N. H. (2006) 'ISRIC-WISE derived soil properties on a 5 by 5 arc-minutes global grid', Report 2006/02, ISRIC – World Soil Information, Wageningen, Netherlands, available through www.isric.org

Berger, M. and Finkbeiner, M. (2010) 'Water footprinting: How to address water use in life cycle assessment?', *Sustainability*, vol 2, pp919–944

Brown, S., Schreier, H. and Lavkulich, L. M. (2009) 'Incorporating virtual water into water management: A British Columbia example', *Water Resources Management*, vol 23, no 13, pp2681–2696

Bulsink, F., Hoekstra, A. Y. and Booij, M. J. (2010) 'The water footprint of Indonesian provinces related to the consumption of crop products', *Hydrology and Earth System Sciences*, vol 14, no 1, pp119–128

Canadian Council of Ministers of the Environment (2010) 'Canadian water quality guidelines for the protection of aquatic life', Canadian Environmental Quality Guidelines, Canadian Council of Ministers of the Environment, Winnipeg, Canada, http://ceqg-rcqe.ccme.ca

CBD (Convention on Biological Diversity) (2002) 'Global strategy for plant conservation', CBD, Montreal, Canada, www.cbd.int

Chahed, J., Hamdane, A. and Besbes, M. (2008) 'A comprehensive water balance of Tunisia: Blue water, green water and virtual water', *Water International*, vol 33, no 4, pp415–424

Chapagain, A. K. and Hoekstra, A. Y. (2003) 'Virtual water flows between nations in relation to trade in livestock and livestock products', Value of Water Research Report Series No.13, UNESCO-IHE, Delft, Netherlands, www.waterfootprint.org/Reports/Report13.pdf

Chapagain, A. K. and Hoekstra, A. Y. (2004) 'Water footprints of nations', Value of Water Research Report Series No.16, UNESCO-IHE, Delft, Netherlands, www.waterfootprint.org/Reports/Report16Vol1.pdf

Chapagain, A. K., and Hoekstra, A. Y. (2007) 'The water footprint of coffee and tea consumption in the Netherlands', *Ecological Economics*, vol 64, no 1, pp109–118

Chapagain, A. K. and Hoekstra, A. Y. (2008) 'The global component of freshwater demand and supply: An assessment of virtual water flows between nations as a result of trade in agricultural and industrial products', *Water International*, vol 33, no 1, pp19–32

Chapagain, A. K. and Hoekstra, A. Y. (2010) 'The green, blue and grey water footprint of rice from both a production and consumption perspective', Value of Water Research Report Series No.40, UNESCO-IHE, Delft, Netherlands, www.waterfootprint.org/Reports/Report40-WaterFootprintRice.pdf

Chapagain, A. K. and Orr, S. (2008) *UK Water Footprint: The Impact of the UK's Food and Fibre Consumption on Global Water Resources*, WWF-UK, Godalming

Chapagain, A. K., and Orr, S. (2009) 'An improved water footprint methodology linking global consumption to local water resources: A case of Spanish tomatoes', *Journal of Environmental Management*, vol 90, pp1219–1228

Chapagain, A. K. and Orr, S. (2010) 'Water footprint of Nestlé's "Bitesize Shredded Wheat": A pilot study to account and analyse the water footprints of Bitesize Shredded Wheat in the context of water availability along its supply chain', WWF-UK, Godalming

Chapagain, A. K., Hoekstra, A. Y. and Savenije, H. H. G. (2006a) 'Water saving through international trade of agricultural products', *Hydrology and Earth System Sciences*, vol 10, no 3, pp455–468

Chapagain, A. K., Hoekstra, A. Y., Savenije, H. H. G. and Gautam, R. (2006b) 'The water footprint of cotton consumption: An assessment of the impact of worldwide consumption of cotton products on the water resources in the cotton producing countries', *Ecological Economics*, vol 60, no 1, pp186–203

Chinese Ministry of Environmental Protection (2002) 'Environmental quality standard for surface water', Ministry of Environmental Protection, The People's Republic of China, http://english.mep.gov.cn/standards_reports/standards/water_environment/quality_standard/200710/t20071024_111792.htm

Clark, G. M., Mueller, D. K., Mast, M. A. (2000) 'Nutrient concentrations and yields in undeveloped stream basins of the United States', *Journal of the American Water Resources Association*, vol 36, no 4, pp849–860

CONAMA (Conselho Nacional do Meio Ambiente) (2005) *Brazilian Water Quality Standards for Rivers*, The National Council of the Environment, Brazilian Ministry of the Environment

Crommentuijn, T., Sijm, D., de Bruijn, J., van den Hoop, M., van Leeuwen, K. and van de Plassche, E. (2000) 'Maximum permissible and negligible concentrations for metals and metalloids in the Netherlands, taking into account background concentrations', *Journal of Environmental Management*, vol 60, pp121–143

CropLife Foundation (2006) *National Pesticide Use Database 2002*, CropLife Foundation, Washington, DC, www.croplifefoundation.org/cpri_npud2002.htm

Dabrowski, J. M., Murray, K., Ashton, P. J. and Leaner, J. J. (2009) 'Agricultural impacts on water quality and implications for virtual water trading decisions', *Ecological Economics*, vol 68, no 4, pp1074–1082

Dastane, N. G. (1978) 'Effective rainfall in irrigated agriculture', Irrigation and Drainage Paper No 25, Food and Agriculture Organization, Rome, www.fao.org/docrep/X5560E/x5560e00.htm#Contents

Dietzenbacher, E. and Velazquez, E. (2007) 'Analysing Andalusian virtual water trade in an input-output framework', *Regional Studies*, vol 41, no 2, pp185–196

Dominguez-Faus, R., Powers, S. E., Burken, J. G. and Alvarez, P. J. (2009) 'The water footprint of biofuels: A drink or drive issue?', *Environmental Science & Technology*, vol 43, no 9, pp3005–3010

Dyson, M., Bergkamp, G. and Scanlon, J. (eds) (2003) *Flow: The Essentials of Environmental Flows*, IUCN, Gland, Switzerland

Ecoinvent (2010) *Ecoinvent Data v2.2*, Ecoinvent Centre, Switzerland, www.ecoinvent.org

Elkington, J. (1997) *Cannibals with Forks: The Triple Bottom Line of 21st Century Business*, Capstone, Oxford

Ene, S. A. and Teodosiu, C. (2009) 'Water footprint and challenges for its application to integrated water resources management in Romania', *Environmental Engineering and Management Journal*, vol 8, no 6, pp1461–1469

Environment Agency (2007) 'Towards water neutrality in the Thames Gateway', summary report, science report SC060100/SR3, Environment Agency, Bristol

EPA (Environmental Protection Agency) (2005) 'List of drinking water contaminants: Ground water and drinking water', US Environmental Protection Agency, www.epa.gov/safewater/mcl.html#1

EPA (2010a) 'Overview of impaired waters and total maximum daily loads program', US Environmental Protection Agency, www.epa.gov/owow/tmdl/intro.html

EPA (2010b) 'National recommended water quality criteria', US Environmental Protection Agency, www.epa.gov/waterscience/criteria/wqctable/index.html#nonpriority

Ercin, A. E., Aldaya, M. M. and Hoekstra, A. Y. (2009) 'A pilot in corporate water footprint accounting and impact assessment: The water footprint of a sugar-containing carbonated beverage', Value of Water Research Report Series No 39, UNESCO-IHE, Delft, Netherlands, www.waterfootprint.org/Reports/Report39-WaterFootprintCarbonatedBeverage.pdf

EU (European Union) (2000) 'Directive 2000/60/EC of the European Parliament and of the Council of 23 October 2000 establishing a framework for Community action in the field of water policy', EU, http://eur-lex.europa.eu/LexUriServ/LexUriServ.do?uri=CONSLEG:2000L0060:20090113:EN:PDF

EU (2006) 'Directive 2006/44/EC of the European Parliament and of the Council of 6 September 2006 on the quality of fresh waters needing protection or improvement in order to support fish life', EU, Brussels, http://eur-lex.europa.eu/LexUriServ/LexUriServ.do?uri=OJ:L:2006:264:0020:0031:EN:pdf

EU (2008) 'Directive 2008/105/EC on environmental quality standards in the field of water policy', EU, http://eur-lex.europa.eu/LexUriServ/LexUriServ.do?uri=OJ:L:2008:348:0084:0097:EN:PDF

Eurostat (2007) *The Use of Plant Protection Products in the European Union: Data 1992–2003*, Eurostat Statistical Books, European Commission, http://epp.eurostat.ec.europa.eu/cache/ITY_OFFPUB/KS-76-06-669/EN/KS-76-06-669-EN.PDF

Falkenmark, M. (1989) 'The massive water scarcity now threatening Africa: Why isn't it being addressed?', *Ambio*, vol 18, no 2, pp112–118

Falkenmark, M. (2003) 'Freshwater as shared between society and ecosystems: from divided approaches to integrated challenges', *Philosophical Transaction of the Royal Society of London B*, vol 358, no 1440, pp2037–2049

Falkenmark, M. and Lindh, G. (1974) 'How can we cope with the water resources situation by the year 2015?', *Ambio*, vol 3, nos 3–4, pp114–122

Falkenmark, M. and Rockström, J. (2004) *Balancing Water for Humans and Nature: The New Approach in Ecohydrology*, Earthscan, London

FAO (Food and Agriculture Organization) (2003) Technical conversion factors for agricultural commodities, FAO, Rome, www.fao.org/fileadmin/templates/ess/documents/methodology/tcf.pdf

FAO (2005) 'New LocClim, Local Climate Estimator CD-ROM', FAO, Rome, www.fao.org/nr/climpag/pub/en3_051002_en.asp

FAO (2010a) 'CLIMWAT 2.0 database', FAO, Rome, www.fao.org/nr/water/infores_databases_climwat.html

FAO (2010b) 'CROPWAT 8.0 model', FAO, Rome, www.fao.org/nr/water/infores_databases_cropwat.html.

FAO (2010c) 'FertiStat database', FAO, Rome, www.fao.org/ag/agl/fertistat

FAO (2010d) 'FAOSTAT database', FAO, Rome, http://faostat.fao.org

FAO (2010e) 'AQUACROP 3.1', FAO, Rome, www.fao.org/nr/water/aquacrop.html

FAO (2010f) 'Global Information and Early Warning System (GIEWS)', FAO, Rome, www.fao.org/giews/countrybrief/index.jsp

FAO (2010g) 'Global map of monthly reference evapotranspiration and precipitation – at 10 arc minutes', GeoNetwork grid database, www.fao.org/geonetwork/srv/en

FAO (2010h) 'Global map maximum soil moisture – at 5 arc minutes', GeoNetwork grid database, www.fao.org/geonetwork/srv/en.

Galan-del-Castillo, E. and Velazquez, E. (2010) 'From water to energy: The virtual water content and water footprint of biofuel consumption in Spain', *Energy Policy*, vol 38, no 3, pp1345–1352

Galloway, J. N., Burke, M., Bradford, G. E., Naylor, R., Falcon, W., Chapagain, A. K., Gaskell, J. C., McCullough, E., Mooney, H. A., Oleson, K. L. L., Steinfeld, H., Wassenaar, T. and Smil, V. (2007) 'International trade in meat: The tip of the pork chop', *Ambio*, vol 36, no 8, pp622–629

Garrido, A., Llamas, M. R., Varela-Ortega, C., Novo, P., Rodríguez-Casado, R. and Aldaya, M. M. (2010) *Water Footprint and Virtual Water Trade in Spain*, Springer, New York, NY

Gerbens-Leenes, P. W. and Hoekstra, A. Y. (2009) 'The water footprint of sweeteners and bio-ethanol from sugar cane, sugar beet and maize', Value of Water Research Report Series No 38, UNESCO-IHE, Delft, Netherlands, www.waterfootprint.org/Reports/Report38-WaterFootprint-sweeteners-ethanol.pdf

Gerbens-Leenes, P. W. and Hoekstra, A. Y. (2010) 'Burning water: The water footprint of biofuel-based transport', Value of Water Research Report Series No 44, UNESCO-IHE, Delft, Netherlands, www.waterfootprint.org/Reports/Report44-BurningWater-WaterFootprintTransport.pdf

Gerbens-Leenes, P. W., Hoekstra, A. Y. and Van der Meer. T. H. (2009a) 'The water footprint of energy from biomass: A quantitative assessment and consequences of an increasing share of bio-energy in energy supply', *Ecological Economics*, vol 68, no 4, pp1052–1060

Gerbens-Leenes, W., Hoekstra, A. Y. and Van der Meer, T. H. (2009b) 'The water footprint of bioenergy', *Proceedings of the National Academy of Sciences*, vol 106, no 25, pp10219–10223

Gerbens-Leenes, W., Hoekstra, A. Y. and Van der Meer, T. H. (2009c) 'A global estimate of the water footprint of *Jatropha curcas* under limited data availability', *Proceedings of the National Academy of Sciences*, vol 106, no 40, pE113

Gleick, P. H. (ed) (1993) *Water in Crisis: A Guide to the World's Fresh Water Resources*, Oxford University Press, Oxford

Gleick, P. H. (2010) 'Water conflict chronology', www.worldwater.org/conflict

GWP (Global Water Partnership) (2000) 'Integrated water resources management', TAC Background Papers No 4, GWP, Stockholm

GWP and INBO (International Network of Basin Organizations) (2009) *A Handbook for Integrated Water Resources Management in Basins*, GWP, Stockholm, and INBO, Paris

Heffer, P. (2009) 'Assessment of fertilizer use by crop at the global level', International Fertilizer Industry Association, Paris, www.fertilizer.org/ifa/Home-Page/LIBRARY/Publication-database.html/Assessment-of-Fertilizer-Use-by-Crop-at-the-Global-Level-2006-07-2007-08.html2

Herendeen, R. A. (2004) 'Energy analysis and EMERGY analysis: A comparison', *Ecological Modelling*, vol 178, pp227–237.

Hoekstra, A. Y. (ed) (2003) 'Virtual water trade: Proceedings of the International Expert Meeting on Virtual Water Trade', 12–13 December 2002, Value of Water Research Report Series No 12, UNESCO-IHE, Delft, Netherlands, www.waterfootprint.org/Reports/Report12.pdf

Hoekstra, A. Y. (2006) 'The global dimension of water governance: Nine reasons for global arrangements in order to cope with local water problems', Value of Water Research Report Series No 20, UNESCO-IHE, Delft, Netherlands, www.waterfootprint.org/Reports/Report_20_Global_Water_Governance.pdf

Hoekstra, A. Y. (2008a) 'Water neutral: Reducing and offsetting the impacts of water footprints', Value of Water Research Report Series No 28, UNESCO-IHE, Delft, Netherlands, www.waterfootprint.org/Reports/Report28-WaterNeutral.pdf

Hoekstra, A. Y. (2008b) 'The relation between international trade and water resources management', in K. P. Gallagher (ed) *Handbook on Trade and the Environment*, Edward Elgar Publishing, Cheltenham, pp116–125

Hoekstra, A. Y. (2008c) 'The water footprint of food', in J. Förare (ed) *Water For Food*, The Swedish Research Council for Environment, Agricultural Sciences and Spatial Planning, Stockholm,, pp49–60

Hoekstra, A. Y. (2009) 'Human appropriation of natural capital: A comparison of ecological footprint and water footprint analysis', *Ecological Economics*, vol 68, no 7, pp1963–1974

Hoekstra, A. Y. (2010a) 'The relation between international trade and freshwater scarcity', Working Paper ERSD-2010-05, January 2010, World Trade Organization, Geneva

Hoekstra, A. Y. (2010b) 'The water footprint of animal products', in J. D'Silva and J. Webster (eds) *The Meat Crisis: Developing More Sustainable Production and Consumption*, Earthscan, London, pp22–33

Hoekstra, A. Y. and Chapagain, A. K. (2007a) 'Water footprints of nations: Water use by people as a function of their consumption pattern', *Water Resources Management*, vol 21, no 1, pp35–48

Hoekstra, A. Y. and Chapagain, A. K. (2007b) 'The water footprints of Morocco and the Netherlands: Global water use as a result of domestic consumption of agricultural commodities', *Ecological Economics*, vol 64, no 1, pp143–151

Hoekstra, A. Y. and Chapagain, A. K. (2008) *Globalization of Water: Sharing the Planet's Freshwater Resources*, Blackwell Publishing, Oxford

Hoekstra, A. Y. and Hung, P. Q. (2002) 'Virtual water trade: A quantification of virtual water flows between nations in relation to international crop trade', Value of Water Research Report Series No 11, UNESCO-IHE, Delft, Netherlands, www.waterfootprint.org/Reports/Report11.pdf

Hoekstra, A. Y. and Hung, P. Q. (2005) 'Globalisation of water resources: International virtual water flows in relation to crop trade', *Global Environmental Change*, vol 15, no 1, pp45–56

Hoekstra, A. Y., Chapagain, A. K., Aldaya, M. M. and Mekonnen, M. M. (2009a) *Water Footprint Manual: State of the Art 2009*, Water Footprint Network, Enschede, the Netherlands, www.waterfootprint.org/downloads/WaterFootprintManual2009.pdf

Hoekstra, A. Y., Gerbens-Leenes, W. and Van der Meer, T. H. (2009b) 'Water footprint accounting, impact assessment, and life-cycle assessment', *Proceedings of the National Academy of Sciences*, vol 106, no 40, pE114

Hoekstra, A. Y., Gerbens-Leenes, W. and Van der Meer, T. H. (2009c) 'The water footprint of *Jatropha curcas* under poor growing conditions', *Proceedings of the National Academy of Sciences*, vol 106, no 42, pE119

Hubacek, K., Guan, D. B., Barrett, J. and Wiedmann, T. (2009) 'Environmental implications of urbanization and lifestyle change in China: Ecological and water footprints', *Journal of Cleaner Production*, vol 17, no 14, pp1241–1248

Humbert, S., Loerincik, Y., Rossi, V., Margnia, M. and Jolliet, O. (2009) 'Life cycle assessment of spray dried soluble coffee and comparison with alternatives (drip filter and capsule espresso)', *Journal of Cleaner Production*, vol 17, no 15, pp1351–1358

IFA (International Fertilizer Industry Association) (2009) 'IFA data', IFA, www.fertilizer.org/ifa/ifadata/search

IFC, LimnoTech, Jain Irrigation Systems and TNC (2010) *Water Footprint Assessments: Dehydrated Onion Products, Micro-irrigation Systems* – Jain Irrigation Systems Ltd, International Finance Corporation, Washington, DC

IPCC (Intergovernmental Panel on Climate Change) (2006) '2006 IPCC guidelines for national greenhouse gas inventories', IPCC, www.ipcc-nggip.iges.or.jp

Japanese Ministry of the Environment (2010) 'Environmental quality standards for water pollution', Ministry of the Environment, Government of Japan, www.env.go.jp/en/water

Jongschaap, R. E. E., Blesgraaf, R. A. R., Bogaard, T. A., Van Loo, E. N. and Savenije, H. H. G. (2009) 'The water footprint of bioenergy from *Jatropha curcas* L.', *Proceedings of the National Academy of Sciences*, vol 106, no 35, ppE92–E92

Kampman, D. A., Hoekstra, A. Y. and Krol, M. S. (2008) 'The water footprint of India', Value of Water Research Report Series No 32, UNESCO-IHE, Delft, Netherlands

Koehler, A. (2008) 'Water use in LCA: Managing the planet's freshwater resources', *International Journal of Life Cycle Assessment*, vol 13, no 6, pp451–455

Kuiper, J., Zarate, E, and Aldaya, M. (2010) 'Water footprint assessment, policy and practical measures in a specific geographical setting', A study in collaboration with the UNEP Division of Technology, Industry and Economics, Water Footprint Network, Enschede, Netherlands

Kumar, V. and Jain, S. K. (2007) 'Status of virtual water trade from India', *Current Science*, vol 93, pp1093–1099

LAWA-AO (2007) 'Monitoring framework design, Part B, Valuation bases and methods descriptions: Background and guidance values for physico-chemical components', www.vsvi-sachsen.de/Beitr%E4ge%20aus%20unseren%20Veranst/17.09.2008%20Tausalz%20Recht%20RAKONArbeitspapierII_Stand_07_03_2007.pdf

Levinson, M., Lee, E., Chung, J., Huttner, M., Danely, C., McKnight, C. and Langlois, A. (2008) *Watching Water: A Guide to Evaluating Corporate Risks in a Thirsty World*, J. P. Morgan, New York, NY

Liu, J. and Savenije, H. H. G. (2008) 'Food consumption patterns and their effect on water requirement in China', *Hydrology and Earth System Sciences*, vol 12, no 3, pp887–898.

Liu, J. G., Williams, J. R., Zehnder, A. J. B. and Yang, H. (2007) 'GEPIC: Modelling wheat yield and crop water productivity with high resolution on a global scale', *Agricultural Systems*, vol 94, no 2, pp478–493

Liu, J., Zehnder, A. J. B. and Yang, H. (2009) 'Global consumptive water use for crop production: The importance of green water and virtual water', *Water Resources Research*, vol 45, pW05428

Ma, J., Hoekstra, A. Y., Wang, H., Chapagain, A. K. and Wang, D. (2006) 'Virtual versus real water transfers within China', *Philosophical Transactions of the Royal Society B: Biological Sciences*, vol 361, no 1469, pp835–842

MacDonald, D. D., Berger, T., Wood, K., Brown, J., Johnsen, T., Haines, M. L., Brydges, K., MacDonald, M. J., Smith, S. L. and Shaw, D. P. (2000) *Compendium of Environmental Quality Benchmarks*, MacDonald Environmental Sciences, Nanaimo, British Columbia, www.pyr.ec.gc.ca/georgiabasin/reports/Environmental%20Benchmarks/GB-99-01_E.pdf

Maes, W. H., Achten, W. M. J. and Muys, B. (2009) 'Use of inadequate data and methodological errors lead to an overestimation of the water footprint of Jatropha curcas', *Proceedings of the National Academy of Sciences*, vol 106, no 34, ppE91–E91

MAPA (2001) *Calendario de siembra, recolección y comercialización, años 1996–1998*, Spanish Ministry of Agriculture, Madrid

MARM (2009) *Agro-alimentary Statistics Yearbook*, Spanish Ministry of the Environment and Rural and Marine Affairs, www.mapa.es/es/estadistica/pags/anuario/introduccion.htm

Mekonnen, M. M. and Hoekstra, A. Y. (2010a) 'A global and high-resolution assessment of the green, blue and grey water footprint of wheat', *Hydrology and Earth System Sciences*, vol 14, pp1259–1276

Mekonnen, M. M. and Hoekstra, A. Y. (2010b) 'Mitigating the water footprint of export cut flowers from the Lake Naivasha Basin, Kenya', Value of Water Research Report Series No 45, UNESCO-IHE, Delft, Netherlands, www.waterfootprint.org/Reports/Report45-WaterFootprint-Flowers-Kenya.pdf

Milà i Canals, L., Chenoweth, J., Chapagain, A., Orr, S., Antón, A. and Clift, R. (2009) 'Assessing freshwater use impacts in LCA: Part I – inventory modelling and characterisation factors for the main impact pathways', *Journal of Life Cycle Assessment*, vol 14, no 1, pp28–42

Mitchell, T. D. and Jones, P. D. (2005) 'An improved method of constructing a database of monthly climate observations and associated high-resolution grids', *International Journal of Climatology*, vol 25, pp693–712, http://cru.csi.cgiar.org/continent_selection.asp

Monfreda, C., Ramankutty, N. and Foley, J. A. (2008) 'Farming the planet: 2. Geographic distribution of crop areas, yields, physiological types, and net primary production in the year 2000', *Global Biogeochemical Cycles*, vol 22, no 1, pGB1022, www.geog.mcgill.ca/landuse/pub/Data/175crops2000

Morrison, J., Morkawa, M., Murphy, M. and Schulte, P. (2009) *Water Scarcity and Climate Change: Growing Risks for Business and Investors*, CERES, Boston, MA, www.ceres.org/Document.Doc?id=406

Morrison, J., Schulte, P. and Schenck, R. (2010) *Corporate Water Accounting: An Analysis of Methods and Tools for Measuring Water Use and its Impacts*, United Nations Global Compact, New York, NY, www.pacinst.org/reports/corporate_water_accounting_analysis/corporate_water_accounting_analysis.pdf

NASS (2009) *Agricultural Chemical Use Database*, National Agricultural Statistics Service, www.pestmanagement.info/nass

Nazer, D. W., Siebel, M. A., Van der Zaag, P., Mimi, Z. and Gijzen, H. J. (2008) 'Water footprint of the Palestinians in the West Bank', *Journal of the American Water Resources Association*, vol 44, no 2, pp449–458

NCDC (National Climatic Data Center) (2009) Global surface summary of the day, NCDC, www.ncdc.noaa.gov/cgi-bin/res40.pl?page=gsod.html, data available from ftp://ftp.ncdc.noaa.gov/pub/data/gsod

Noss, R. F. and Cooperrider, A. Y. (1994) *Saving Nature's Legacy: Protecting and Restoring Biodiversity*, Island Press, Washington, DC

Novo, P., Garrido, A. and Varela-Ortega, C. (2009) 'Are virtual water "flows" in Spanish grain trade consistent with relative water scarcity?', *Ecological Economics*, vol 68, no 5, pp1454–1464

Odum, H. T. (1996) *Environmental Accounting: Emergy and Environmental Decision Making*, Wiley, New York, NY

Official State Gazette (2008) 'Approval of the water planning instruction', Ministry of the Environment and Rural and Marine Affairs, Official State Gazette 229, Madrid, Spain, 22 September 2008, www.boe.es/boe/dias/2008/09/22/pdfs/A38472-38582.pdf

Oregon State University (2010) 'The transboundary freshwater dispute database', Oregon State University, Department of Geosciences, Corvallis, OR, www.transboundarywaters.orst.edu/database

Pegram, G., Orr, S. and Williams, C. (2009) *Investigating Shared Risk in Water: Corporate Engagement with the Public Policy Process*, WWF, Godalming

Perry, C. (2007) 'Efficient irrigation; Inefficient communication; Flawed recommendations', *Irrigation and Drainage*, vol 56, no 4, pp367–378

Pfister, S. and Hellweg, S. (2009) 'The water "shoesize" vs. footprint of bioenergy', *Proceedings of the National Academy of Sciences*, vol 106, no 35, ppE93–E94

Pfister, S., Koehler, A. and Hellweg, S. (2009) 'Assessing the environmental impacts of freshwater consumption in LCA', *Environmental Science and Technology*, vol 43, pp4098–4104

Poff, N. L., Richter, B. D., Aarthington, A. H., Bunn, S. E., Naiman, R. J., Kendy, E., Acreman, M., Apse, C., Bledsoe, B. P., Freeman, M. C., Henriksen, J., Jacobson, R. B., Kennen, J. G., Merritt, D. M., O'Keeffe, J. H., Olden, J. D., Rogers, K., Tharme, R. E. and Warner, A. (2010) 'The ecological limits of hydrologic alteration (ELOHA): A new framework for developing regional environmental flow standards', *Freshwater Biology*, vol 55, no 1, pp147–170

Portmann, F., Siebert, S., Bauer, C. and Döll, P. (2008) 'Global data set of monthly growing areas of 26 irrigated crops', Frankfurt Hydrology Paper 06, Institute of Physical Geography, University of Frankfurt, Frankfurt am Main, www.geo.uni-frankfurt.de/ipg/ag/dl/forschung/MIRCA/index.html

Portmann, F. T., Siebert, S. and Döll P. (2010) 'MIRCA2000 – Global monthly irrigated and rainfed crop areas around the year 2000: A new high-resolution data set for agricultural and hydrological modeling', *Global Biogeochemical Cycles*, vol 24, GB1011

Postel, S. L., Daily, G. C. and Ehrlich, P. R. (1996) 'Human appropriation of renewable fresh water', *Science*, vol 271, pp785–788

Raskin, P. D., Hansen, E. and Margolis, R. M. (1996) 'Water and sustainability: global patterns and long-range problems', *Natural Resources Forum*, vol 20, no 1, pp1–5

Rebitzer, G., Ekvall, T., Frischknecht, R., Hunkeler, D., Norris, G., Rydberg, T., Schmidt, W. P., Suh, S., Weidema, B. P. and Pennington, D. W. (2004) 'Life cycle assessment Part 1: Framework, goal and scope definition, inventory analysis, and applications', *Environment International*, vol 30, pp701–720

Rees, W. E. (1992) 'Ecological footprints and appropriated carrying capacity: What urban economics leaves out', *Environment and Urbanization*, vol 4, no 2, pp121–130

Rees, W. E. (1996) 'Revisiting carrying capacity: Area-based indicators of sustainability', *Population and Environment*, vol 17, no 3, pp195–215

Rees, W. E. and Wackernagel, M. (1994) 'Ecological footprints and appropriated carrying capacity: Measuring the natural capital requirements of the human economy', in A. M. Jansson, M. Hammer, C. Folke and R. Costanza (eds) *Investing in Natural Capital: The Ecological Economics Approach to Sustainability*, Island Press, Washington, DC, pp362–390

Richter, B. D. (2010) 'Re-thinking environmental flows: From allocations and reserves to sustainability boundaries', *River Research and Applications*, vol 26, no 8, pp1052–1063

Ridoutt, B. G. and Pfister, S. (2010) 'A revised approach to water footprinting to make transparent the impacts of consumption and production on global freshwater scarcity', *Global Environmental Change*, vol 20, no 1, pp113–120

Ridoutt, B. G., Eady, S. J., Sellahewa, J. Simons, L. and Bektash, R. (2009) 'Water footprinting at the product brand level: case study and future challenges', *Journal of Cleaner Production*, vol 17, no 13, pp1228–1235

Ridoutt, B. G., Juliano, P., Sanguansri, P. and Sellahewa, J. (2010) 'The water footprint of food waste: Case study of fresh mango in Australia', *Journal of Cleaner Production*, vol 18, nos 16–17, pp1714–1721

Rockström, J. (2001) 'Green water security for the food makers of tomorrow: Windows of opportunity in drought-prone savannahs', *Water Science and Technology*, vol 43, no 4, pp71–78

Romaguera, M., Hoekstra, A. Y., Su, Z., Krol, M. S. and Salama, M. S. (2010) 'Potential of using remote sensing techniques for global assessment of water footprint of crops', *Remote Sensing*, vol 2, no 4, pp1177–1196

SABMiller and WWF-UK (2009) *Water Footprinting: Identifying and Addressing Water Risks in the Value Chain,* SABMiller, Woking and WWF-UK, Goldalming

SABMiller, GTZ and WWF (2010) *Water Futures: Working Together for a Secure Water Future*, SABMiller, Woking and WWF-UK, Goldalming

Safire, W. (2008) 'On language: Footprint', *New York Times*, 17 February 2008

Savenije, H. H. G. (2000) 'Water scarcity indicators: The deception of the numbers', *Physics and Chemistry of the Earth*, vol 25, no 3, pp199–204

Siebert, S. and Döll, P. (2010) 'Quantifying blue and green virtual water contents in global crop production as well as potential production losses without irrigation', *Journal of Hydrology*, vol 384, no 3–4, pp198–207

Siebert, S., Döll, P., Feick, S., Hoogeveen, J. and Frenken, K. (2007) 'Global map of irrigation areas, version 4.0.1', Johann Wolfgang Goethe University, Frankfurt am Main, and FAO, Rome, www.fao.org/nr/water/aquastat/irrigationmap/index10.stm

Smakhtin, V., Revenga, C. and Döll, P. (2004) 'A pilot global assessment of environmental water requirements and scarcity', *Water International*, vol 29, no 3, pp307–317

Smith, M. (1992) 'CROPWAT – A computer program for irrigation planning and management', Irrigation and Drainage Paper 46, FAO, Rome

Smith, R. A., Alexander, R. and Schwarz, G. E. (2003) 'Natural background concentrations of nutrients in streams and rivers of the conterminous United States', *Environmental Science and Technology*, vol 37, no 14, pp3039–3047

Sonnenberg, A., Chapagain, A., Geiger, M. and August, D. (2009) *Der Wasser-Fußabdruck Deutschlands: Woher stammt das Wasser, das in unseren Lebensmitteln steckt?*, WWF Deutschland, Frankfurt

South African Department of Water Affairs and Forestry (1996) *South African Water Quality Guidelines*, vol 7, Aquatic Ecosystems, Department of Water Affairs and Forestry

Svancara, L. K., Brannon, R., Scott, J. M., Groves, C. R., Noss, R. F. and Pressey, R. L. (2005) 'Policy-driven versus evidence-based conservation: A review of political targets and biological needs', *BioScience*, vol 55, no 11, pp989–995

TCCC and TNC (The Coca-Cola Company and The Nature Conservancy) (2010) *Product Water Footprint Assessments: Practical Application in Corporate Water Stewardship*, TCCC, Atlanta, and TNC, Arlington

UKTAG (UK Technical Advisory Group) (2008) 'UK environmental standards and conditions (Phase 1)', UK Technical Advisory Group on the Water Framework Directive, www.wfduk.org/UK_Environmental_Standards/ES_Phase1_final_report

UN (United Nations) (1948) *Universal Declaration of Human Rights*, UN General Assembly, Resolution 217 A (III) of 10 December 1948, Paris

UN (2010a) *Trends in Sustainable Development: Towards Sustainable Consumption and Production*, UN, New York, NY, www.un.org/esa/dsd/resources/res_pdfs/publications/trends/trends_sustainable_consumption_production/Trends_in_sustainable_consumption_and_production.pdf

UN (2010b) 'The human right to water and sanitation', UN General Assembly, 64th session, Agenda item 48, UN, New York, NY

UNEP (United Nations Environment Programme) (2009) 'GEMSTAT: Global water quality data and statistics', Global Environment Monitoring System, UNEP, Nairobi, Kenya, www.gemstat.org

UNESCO (United Nations Educational, Scientific and Cultural Organization) (2009) *IWRM Guidelines at River Basin Level, Part I: Principles*, UNESCO, Paris

USDA (United States Department of Agriculture) (1994) 'The major world crop areas and climatic profiles', *Agricultural Handbook No 664*, World Agricultural Outlook Board, USDA, www.usda.gov/oce/weather/pubs/Other/MWCACP/MajorWorldCropAreas.pdf

Van der Leeden, F., Troise, F. L. and Todd, D. K. (1990) *The Water Encyclopedia*, second edition, CRC Press, Boca Raton,FL

Van Lienden, A. R., Gerbens-Leenes, P. W., Hoekstra, A. Y. and Van der Meer, T. H. (2010) 'Biofuel scenarios in a water perspective: The global blue and green water footprint of road transport in 2030', Value of Water Research Report Series No 43, UNESCO-IHE, Delft, Netherlands, www.waterfootprint.org/Reports/Report43-WaterFootprint-BiofuelScenarios.pdf

Van Oel, P. R. and Hoekstra, A. Y. (2010) 'The green and blue water footprint of paper products: Methodological considerations and quantification', Value of Water Research Report Series No 46, UNESCO-IHE, Delft, Netherlands, www.waterfootprint.org/Reports/Report46-WaterFootprintPaper

Van Oel, P. R., Mekonnen M. M. and Hoekstra, A. Y. (2008) 'The external water footprint of the Netherlands: Quantification and impact assessment', Value of Water Research Report Series No 33, UNESCO-IHE, Delft, Netherlands, www.waterfootprint.org/Reports/Report33-ExternalWaterFootprintNetherlands.pdf

Van Oel, P. R., Mekonnen M. M. and Hoekstra, A. Y. (2009) 'The external water footprint of the Netherlands: Geographically-explicit quantification and impact assessment', *Ecological Economics*, vol 69, no 1, pp82–92

Verkerk, M. P., Hoekstra, A. Y. and Gerbens-Leenes, P. W. (2008) 'Global water governance: Conceptual design of global institutional arrangements', Value of Water Research Report Series No 26, UNESCO-IHE, Delft, Netherlands, www.waterfootprint.org/Reports/Report26-Verkerk-et-al-2008GlobalWaterGovernance.pdf

Verma, S., Kampman, D. A., Van der Zaag, P. and Hoekstra, A. Y. (2009) 'Going against the flow: A critical analysis of inter-state virtual water trade in the context of India's National River Linking Programme', *Physics and Chemistry of the Earth*, vol 34, pp261–269

Wackernagel, M. and Rees, W. (1996) *Our Ecological Footprint: Reducing Human Impact on the Earth*, New Society Publishers, Gabriola Island, BC, Canada

Wang, H. R. and Wang, Y. (2009) 'An input-output analysis of virtual water uses of the three economic sectors in Beijing', *Water International*, vol 34, no 4, pp451–467

Water Neutral (2002) 'Get water neutral!', brochure distributed among delegates at the 2002 World Summit on Sustainable Development in Johannesburg, The Water Neutral Foundation, Johannesburg, South Africa.

WCED (World Commission on Environment and Development) (1987) *Our Common Future*, WCED, Oxford University Press, Oxford

Williams, J. R. (1995) 'The EPIC model', in V. P. Singh (ed) *Computer Models of Watershed Hydrology*, Water Resources Publisher, CO, pp909–1000

Williams, J. R., Jones, C.A., Kiniry, J. R. and Spanel, D. A. (1989) 'The EPIC crop growth-model', *Transactions of the ASAE*, vol 32, no 2, pp497–511

WRI and WBCSD (World Resources Institute and World Business Council for Sustainable Development) (2004) *The Greenhouse Gas Protocol: A Corporate Accounting and Reporting Standard*, revised edition, WRI, Washington, DC andWBCSD, Conches-Geneva, www.ghgprotocol.org/files/ghg-protocol-revised.pdf

WWAP (World Water Assessment Programme) (2009) *The United Nations World Water Development Report 3: Water in a Changing World*, WWAP, UNESCO Publishing, Paris, and Earthscan, London

WWF (2008) *Living Planet Report 2008*, WWF International, Gland, Switzerland.

WWF (2010) *Living Planet Report 2010*, WWF International, Gland, Switzerland.

Yang, H., Zhou, Y. and Liu, J. G. (2009) 'Land and water requirements of biofuel and implications for food supply and the environment in China', *Energy Policy*, vol 37, no 5, pp1876–1885

Yu, Y., Hubacek. K., Feng, K. S. and Guan, D. (2010) 'Assessing regional and global water footprints for the UK', Ecological Economics, vol 69, no 5, pp1140–1147

Zarate, E. (ed) (2010a) 'WFN grey water footprint working group final report: A joint study developed by WFN partners', Water Footprint Network, Enschede, Netherlands

Zarate, E. (ed) (2010b) 'WFN water footprint sustainability assessment working group final report: A joint study developed by WFN partners', Water Footprint Network, Enschede, Netherlands

Zeitoun, M., Allan, J. A. and Mohieldeen, Y. (2010) 'Virtual water "flows" of the Nile Basin, 1998–2004: A first approximation and implications for water security', *Global Environmental Change*, vol 20, no 2, pp229–242

Zhao, X., Chen, B. and Yang, Z. F. (2009) 'National water footprint in an input-output framework: A case study of China 2002', *Ecological Modelling*, vol 220, no 2, pp245–253

Zwart, S. J., Bastiaanssen, W. G. M., De Fraiture, C. and Molden, D. J. (2010) 'A global benchmark map of water productivity for rainfed and irrigated wheat', *Agricultural Water Management*, vol 97, no 10, pp1617–1627

List of Symbols

Symbol	Unit [a]	Explanation
α	–	leaching-run-off fraction, i.e. fraction of applied chemicals reaching freshwater bodies
Abstr	volume/time	volume of water abstraction
Appl	mass/time	application of a chemical (fertilizer or pesticide) per unit of time
AR	mass/area	application rate of a chemical (fertilizer or pesticide) per unit of land
C	mass/time [b]	consumption of a product
c_{act}	mass/volume	actual concentration of a chemical in a water body from which water is abstracted
c_{effl}	mass/volume	concentration of a chemical in an effluent
c_{max}	mass/volume	maximum acceptable concentration of a chemical in a receiving water body
c_{nat}	mass/volume	natural concentration of a chemical in the receiving water body
CWR	length/time	crop water requirement
CWU_{blue}	volume/area	blue crop water use
CWU_{green}	volume/area	green crop water use
E	money/time	total economic value of a product produced in a business unit
Effl	volume/time	volume of effluent (wastewater flow)
EFR	volume/time	environmental flow requirement
ET_a	length/time	adjusted crop evapotranspiration (under actual conditions)
ET_{blue}	length/time	blue water evapotranspiration
ET_c	length/time	crop evapotranspiration (under optimal conditions)
ET_{env}	volume/time	evapotranspiration from land reserved for natural vegetation
ET_{green}	length/time	green water evapotranspiration

Symbol	Unit [a]	Explanation
ET_o	length/time	reference crop evapotranspiration
ET_{unprod}	volume/time	evapotranspiration from land that cannot be made productive in crop production
$f_p[p,i]$	–	product fraction of output product p that is produced from input product i
$f_v[p]$	–	value fraction of output product p
IR	length/time	irrigation requirement
K_c	–	crop coefficient
K_{cb}	–	basal crop coefficient
K_e	–	soil evaporation coefficient
K_s	–	water stress coefficient
L	mass/time	load of a pollutant
L_{crit}	mass/time	critical load of a pollutant
P	mass/time [b]	production quantity of a product
P_{eff}	length/time	effective rainfall
$price$	money/mass	price of a product
R_{act}	volume/time	actual run-off from a catchment
R_{nat}	volume/time	natural run-off from a catchment (without blue water footprint within the catchment)
S_g	volume/time	global water saving through trade in a product
S_n	volume/time	national water saving through trade in a product
T	mass/time [b]	volume of trade in a product
T_e	mass/time [b]	volume of export of a product
T_i	mass/time [b]	volume of import of a product
T_{effl}	temperature	temperature of an effluent
T_{max}	temperature	maximum acceptable temperature for a receiving water body
T_{nat}	temperature	natural temperature of a receiving water body
V_b	volume/time	virtual-water budget of a delineated area (e.g. a nation)
V_e	volume/time	gross virtual-water export from a delineated area (e.g. a nation)
$V_{e,d}$	volume/time	gross virtual-water export insofar concerning export of domestically produced products
$V_{e,r}$	volume/time	gross virtual-water export insofar concerning re-export of imported products
V_i	volume/time	gross virtual-water import into a delineated area (e.g. a nation)

Symbol	Unit [a]	Explanation
$V_{i,net}$	volume/time	net virtual-water import into a delineated area (e.g. a nation)
$w[i]$	mass	quantity of input product *i*
$w[p]$	mass	quantity of output product *p*
WA_{blue}	volume/time	blue water availability
WA_{green}	volume/time	green water availability
WD	%	national virtual-water import dependency
WF_{area}	volume/time	water footprint within a geographically delineated area
$WF_{area,nat}$	volume/time	water footprint within a nation
WF_{bus}	volume/time	water footprint of a business
$\mathrm{WF}_{\mathrm{bus,oper}}$	volume/time	operational water footprint of a business
$WF_{bus,sup}$	volume/time	supply-chain water footprint of a business
WF_{cons}	volume/time	water footprint of a consumer
$WF_{cons,dir}$	volume/time	direct water footprint of a consumer
$WF_{cons,indir}$	volume/time	indirect water footprint of a consumer
$WF_{cons,nat}$	volume/time	water footprint of national consumption
$WF_{cons,nat,dir}$	volume/time	direct water footprint of the consumers in a nation
$WF_{cons,nat,indir}$	volume/time	indirect water footprint of the consumers in a nation
$WF_{cons,nat,ext}$	volume/time	external water footprint of the consumers in a nation
$WF_{cons,nat,int}$	volume/time	internal water footprint of the consumers in a nation
WF_{proc}	volume/time [c]	water footprint of a process
$WF_{proc,blue}$	volume/time [c]	blue water footprint of a process
$WF_{proc,green}$	volume/time [c]	green water footprint of a process
$WF_{proc,grey}$	volume/time [c]	grey water footprint of a process
WF_{prod}	volume/mass [b]	water footprint of a product
WF^{*}_{prod}	volume/mass [b]	average water footprint of a product as available to the consumer or for export
$WFII_{blue}$	–	blue water footprint impact index
$WFII_{green}$	–	green water footprint impact index
$WFII_{grey}$	–	grey water footprint impact index
WPL	–	water pollution level in a catchment area in a specific period within the year
WS_{blue}	–	blue water scarcity in a catchment area in a specific period within the year

Symbol	Unit [a]	Explanation
WS_{green}	–	green water scarcity in a catchment area in a specific period within the year
WSS	%	national water self-sufficiency
Y	mass/area	crop yield

Dimension	Explanation
i	input product
n	nation
n_e	exporting nation
n_i	importing nation
p	(output) product
q	process
s	process step
t	time
u	business unit
x	place / place of origin

[a] The unit of each variable is expressed here in general terms (mass, length, surface, volume, time). In water footprint accounting practice, mass is usually expressed in kg or ton, volume in litres or m^3 and time in day, month or year. Variables like rainfall, evapotranspiration and crop water requirement are usually expressed as mm per day, month or year. Yield and crop water use are usually expressed as ton/ha and m^3/ha respectively. Water quantities are usually expressed as a volume, under the assumption that 1 litre of water is equal to 1 kg. Working with this assumption, mass balances translate in volume balances. Obviously, in reporting numbers it is essential to specify the units used.

[b] A product water footprint is often expressed in terms of water volume per unit of mass; in this case we need to express production, consumption and trade in products in terms of mass/time. A product water footprint, however, can also be expressed in terms of water volume per unit of money; in this case we need to express production, consumption and trade in products in terms of monetary units/time. Other alternative ways to express a product water footprint are for example water volume / piece (for products that are counted per piece rather than weight), water volume / kcal (for food products) or water volume / joule (for electricity or fuels).

[c] A process water footprint is generally expressed in terms of water volume per unit of time. However, through dividing by the amount of product that results from the process (product units/time), a process water footprint also be expressed in terms of water volume per product unit.

Glossary

Ambient water quality standards – The maximum allowable amount of a substance in rivers, lakes or groundwater, given as a concentration. Ambient water quality standards can also refer to other properties of the water, such as temperature or pH. Standards are set to protect against anticipated adverse effects on human health or welfare, wildlife or the functioning of ecosystems.

Blue water – Fresh surface and groundwater, in other words, the water in freshwater lakes, rivers and aquifers.

Blue water availability – Natural run-off (through groundwater and rivers) minus environmental flow requirements. Blue water availability typically varies within the year and also from year to year.

Blue water footprint – Volume of surface and groundwater consumed as a result of the production of a good or service. Consumption refers to the volume of freshwater used and then evaporated or incorporated into a product. It also includes water abstracted from surface or groundwater in a catchment and returned to another catchment or the sea. It is the amount of water abstracted from groundwater or surface water that does not return to the catchment from which it was withdrawn.

Blue water footprint impact index – An aggregated and weighed measure of the environmental impact of a blue water footprint at catchment level. It is based on two inputs: (i) the blue water footprint of a product, consumer or producer specified by catchment and by month; and (ii) the blue water scarcity by catchment and by month. The index is obtained by multiplying the two matrices and then summing the elements of the resultant matrix. The outcome can be interpreted as a blue water footprint weighed according to the blue water scarcity in the places and periods where the various blue water footprint components occur.

Blue water scarcity – The ratio of blue water footprint to blue water availability. Blue water scarcity varies within the year and from year to year.

Business water footprint – See 'water footprint of a business'.

Corporate water footprint – See 'water footprint of a business'.

Critical load – The load of pollutants that will fully consume the assimilation capacity of the receiving water body.

Crop water requirement – The total water needed for evapotranspiration, from planting to harvest for a given crop in a specific climate regime, when adequate soil water is maintained by rainfall and/or irrigation so that it does not limit plant growth and crop yield.

Crop yield – Weight of harvested crop per unit of harvested area.

Dilution factor – The number of times that a polluted effluent volume has to be diluted with ambient water in order to arrive at the maximum acceptable concentration level.

Direct water footprint – The direct water footprint of a consumer or producer (or a group of consumers or producers) refers to the freshwater consumption and pollution that is associated to the water use by the consumer or producer. It is distinct from the indirect water footprint, which refers to the water consumption and pollution that can be associated with the production of the goods and services consumed by the consumer or the inputs used by the producer.

Effective precipitation – The portion of the total precipitation that is retained by the soil so that it is available for crop production.

End-use water footprint of a product – When consumers use a product, there can be a water footprint in the end-use stage. Think about the water pollution that results from the use of soaps in the household. In this case, one can speak about the end-use water footprint of a product. This footprint is strictly spoken not part of the product water footprint, but part of the consumer's water footprint.

Environmental flow requirements – The quantity, quality and timing of water flows required to sustain freshwater and estuarine ecosystems and the human livelihoods and well-being that depend on these ecosystems.

Environmental green water requirement – The quantity of green water from lands that need to be reserved for nature and biodiversity preservation and for human livelihoods that depend on the ecosystems in the natural areas.

Evapotranspiration – Evaporation from the soil and soil surface where crops are grown, including the transpiration of water that actually passes crops.

External water footprint of national consumption – The part of the water footprint of national consumption that falls outside the nation considered. It refers to the appropriation of water resources in other nations for the production of goods and services that are imported into and consumed within the nation considered.

Geographic sustainability – The geographic sustainability of the green, blue and grey water footprints in a catchment or river basin can be assessed based on a number of environmental, social and economic sustainability criteria.

Global water saving through trade – International trade can save freshwater globally if a water-intensive commodity is traded from an area where it is produced with high water productivity (small water footprint) to an area with lower water productivity (large water footprint).

Green water – The precipitation on land that does not run off or recharge the groundwater but is stored in the soil or temporarily stays on top of the soil or vegetation. Eventually, this part of precipitation evaporates or transpires through plants. Green water can be made productive for crop growth (although not all green water can be taken up by crops, because there will always be evaporation from the soil and because not all periods of the year or areas are suitable for crop growth).

Green water availability – The evapotranspiration of rainwater from land minus evapotranspiration from land reserved for natural vegetation and minus evapotranspiration from land that cannot be made productive.

Green water footprint – Volume of rainwater consumed during the production process. This is particularly relevant for agricultural and forestry products (products based on crops or wood), where it refers to the total rainwater evapotranspiration (from fields and plantations) plus the water incorporated into the harvested crop or wood.

Green water footprint impact index – An aggregated and weighed measure of the environmental impact of a green water footprint at catchment level. It is based on two inputs: (i) the green water footprint of a product, consumer or producer specified by catchment and by month; and (ii) the green water scarcity by catchment and by month. The index is obtained by multiplying the two matrices and then summing the elements of the resultant matrix. The outcome can be interpreted as a green water footprint weighed according to the green water scarcity in the places and periods where the various green water footprint components occur.

Green water scarcity – The ratio of green water footprint to green water availability. Green water scarcity varies within the year and from year to year.

Grey water footprint – The grey water footprint of a product is an indicator of freshwater pollution that can be associated with the production of a product over its full supply chain. It is defined as the volume of freshwater that is required to assimilate the load of pollutants based on natural background concentrations and existing ambient water quality standards. It is calculated as the volume of water that is required to dilute pollutants to such an extent that the quality of the water remains above agreed water quality standards.

Grey water footprint impact index – An aggregated and weighed measure of the environmental impact of a grey water footprint at catchment level. It is based on two inputs: (i) the grey water footprint of a product, consumer or producer specified by catchment and by month; and (ii) the water pollution level by catchment and by month. The index is obtained by multiplying the two matrices and then summing the elements of the resultant matrix. The outcome can be interpreted as a grey water footprint weighed according to the water pollution level in the places and periods where the various grey water footprint components occur.

Hotspot – A hotspot is a specific period of the year (such as the dry period) in a specific (sub)catchment in which the water footprint is unsustainable, for example, because it compromises environmental water needs or water quality standards or because the water allocation and use in the catchment is considered unfair and/or economically inefficient.

Indirect water footprint – The indirect water footprint of a consumer or producer refers to the freshwater consumption and pollution 'behind' products being consumed or produced. It is equal to the sum of the water footprints of all products consumed by the consumer or of all (non-water) inputs used by the producer.

Internal water footprint of national consumption – The part of the water footprint of national consumption that falls inside the nation, in other words, the appropriation of domestic water resources for producing goods and services that are consumed domestically.

Irrigation requirement – The quantity of water exclusive of precipitation, in other words, the quantity of irrigation water, required for normal crop production. It includes soil evaporation and some unavoidable losses under the given conditions. It is usually expressed in water-depth units (millimetres) and may be stated in monthly, seasonal or annual terms, or for a crop period.

Maximum acceptable concentration – see 'ambient water quality standards'.

National water footprint – Is the same as what is more accurately called the 'water footprint of national consumption', which is defined as the total amount of fresh water that is used to produce the goods and services consumed by the inhabitants of the nation. Part of this water footprint lies outside the territory of the nation. The term should not be confused with the 'water footprint within a nation', which refers to the total freshwater volume consumed or polluted within the territory of the nation.

National water saving through trade – A nation can preserve its domestic freshwater resources by importing a water-intensive product instead of producing it domestically.

Natural concentration – The natural or background concentration in a receiving water body is the concentration in the water body that would occur if there was no human disturbance in the catchment. (It corresponds to the 'high status' conditions as defined in the EU Water Framework Directive.)

Operational water footprint of a business – The operational (or direct) water footprint of a business is the volume of freshwater consumed or polluted due to its own operations.

Organizational water footprint – See 'water footprint of a business'.

Overhead water footprint – The water footprint of a product consists of two elements: the use of freshwater that can immediately be related to the product and the use of freshwater in overhead activities. The latter element is called the 'overhead water footprint'. The overhead water footprint refers to freshwater use that in the first instance cannot be fully associated with the production of the specific product considered, but refers to freshwater use that associates with supporting activities and materials used in the business, which produces not just this specific product but other products as well. The overhead water footprint of a business has to be distributed over the various business products, which is done based on the relative value per product. The overhead water footprint includes, for example, the freshwater use in the toilets and kitchen of a factory and the freshwater use behind the concrete and steel used in the factory and machineries.

Primary impacts – The term 'primary impacts' is used in the context of assessing the sustainability of a water footprint in a geographic area. Primary impacts refer to the effect of the water footprint in a catchment on water flows and water quality.

Production system – The production system of a product consists of all the sequential process steps applied to produce it. A production system can be a linear chain of processes, it can take the shape of a product tree (many inputs ultimately resulting in one output product) or it may rather look like a complex network of interlinked processes that eventually lead one or more products.

Product tree – See 'production system'.

Return flow – The part of the water withdrawn for an agricultural, industrial or domestic purpose that returns to the groundwater or surface water in the same catchment as where it was abstracted. This water can potentially be withdrawn and used again.

Secondary impacts – The term 'secondary impacts' is used, next to the term 'primary impacts', in the context of assessing the sustainability of a water footprint in a geographic area. Secondary impacts refer to the impacts of a water footprint on ultimate ecological, social and economic values such as biodiversity, human health, welfare and security.

Supply-chain water footprint of a business – The supply-chain (or indirect) water footprint of a business is the volume of freshwater consumed or polluted to produce all the goods and services that form the input of production of a business.

Sustainability criteria – Sustainability criteria are generally categorized into three major themes: environmental, social and economic sustainability.

Virtual-water balance – The virtual-water balance of a geographically delineated area (for example, a nation or catchment area) over a certain time period is defined as the net import of virtual water over this period, which is equal to the gross import of virtual water minus the gross export. A positive virtual-water balance implies net inflow of virtual water to the nation from other nations. A negative balance means net outflow of virtual water.

Virtual-water content – The virtual-water content of a product is the freshwater 'embodied' in the product, not in real sense, but in virtual sense. It refers to the volume of water consumed or polluted for producing the product, measured over its full production chain. If a nation exports/imports such a product, it exports/imports water in virtual form. The 'virtual-water content of a product' is the same as 'the water footprint of a product', but the former refers to the water volume embodied in the product alone, while the latter term refers to that volume, but also to which sort of water is being used and to when and where that water is being used. The water footprint of a product is thus a multi-dimensional indicator, whereas virtual-water content refers to a volume alone.

Virtual-water export – The virtual-water export from a geographically delineated area (for example, a nation or catchment area) is the volume of virtual water associated with the export of goods or services from the area. It is the total volume of freshwater consumed or polluted to produce the products for export.

Virtual-water flow – The virtual-water flow between two geographically delineated areas (for example, two nations) is the volume of virtual water that is being transferred from the one to the another area as a result of product trade.

Virtual-water import – The virtual-water import into a geographically delineated area (for example, a nation or catchment area) is the volume of virtual water associated with the import of goods or services into the area. It is the total volume of freshwater used (in the export areas) to produce the products. Viewed from the perspective of the importing area, this water can be seen as an additional source of water that comes on top of the available water resources within the area itself.

Water abstraction – See 'water withdrawal'.

Water appropriation – This is a term used in the context of water footprint assessment to refer to both the 'consumption' of freshwater for human activities (green and blue water footprint) and the 'pollution' of freshwater by human activities (grey water footprint).

Water consumption – The volume of freshwater used and then evaporated or incorporated into a product. It also includes water abstracted from surface or groundwater in a catchment and returned to another catchment or the sea. It is important to distinguish the term 'water consumption' from the term 'water withdrawal' or 'water abstraction'.

Water footprint – The water footprint is an indicator of freshwater use that looks at both direct and indirect water use of a consumer or producer. The water footprint of an individual, community or business is defined as the total volume of freshwater used to produce the goods and services consumed by the individual or community or produced by the business. Water use is measured in terms of water volumes consumed (evaporated or incorporated into a product) and/or polluted per unit of time. A water footprint can be calculated for a particular product, for any well-defined group of consumers (for example, an individual, family, village, city, province, state or nation) or producers (for example, a public organization, private enterprise or economic sector). The water footprint is a geographically explicit indicator, showing not only volumes of water use and pollution, but also the locations.

Water footprint accounting – The step in water footprint assessment that refers to collecting factual, empirical data on water footprints with a scope and depth as defined earlier.

Water footprint assessment – Water footprint assessment refers to the full range of activities to: (i) quantify and locate the water footprint of a process, product, producer or consumer or to quantify in space and time the water footprint in a specified geographic area; (ii) assess the environmental, social and economic sustainability of this water footprint; and (iii) formulate a response strategy.

Water footprint impact indices – See 'blue/green/grey water footprint impact index'.

Water footprint of a business – The water footprint of a business – which can also be called alternatively corporate or organizational water footprint – is defined as the total volume of freshwater that is used directly and indirectly to run and support a business. The water footprint of a business consists of two components: the direct water use by the producer (for producing/manufacturing or for supporting activities) and the indirect water use (the water use in the producer's supply chain). The 'water footprint of a business' is the same as the total 'water footprint of the business output products'.

Water footprint of a consumer – Is defined as the total volume of freshwater consumed and polluted for the production of the goods and services consumed by the consumer. It is calculated by adding the direct water use by people and their indirect water use. The latter can be found by multiplying all goods and services consumed by their respective water footprint.

Water footprint of national consumption – Is defined as the total amount of fresh water that is used to produce the goods and services consumed by the inhabitants of the nation. The water footprint of national consumption can be assessed in two ways. The bottom-up approach is to consider the sum of all products consumed multiplied with their respective product water footprint. In the top-down approach, the water footprint of national consumption is calculated as the total use of domestic water resources plus the gross virtual-water import minus the gross virtual-water export.

Water footprint of national production – Another term for the 'water footprint within a nation'.

Water footprint of a product – The water footprint of a product (a commodity, good or service) is the total volume of freshwater used to produce the product, summed over the various steps of the production chain. The water footprint of a product refers not only to the total volume of water used; it also refers to where and when the water is used.

Water footprint offsetting – Offsetting the negative impacts of a water footprint is part of water neutrality. Offsetting is a last step, after a prior effort of reducing a water footprint insofar reasonably possible. Compensation can be done by contributing to (for example, by investing in) a more sustainable and equitable use of water in the hydrological units in which the impacts of the remaining water footprint are located.

Water footprint sustainability assessment – The phase in water footprint assessment that aims to evaluate whether a certain water footprint is sustainable from an environmental, social, as well as an economic point of view.

Water footprint within a geographically delineated area – Is defined as the total freshwater consumption and pollution within the boundaries of the area. The area can be for example a hydrological unit such as a catchment area or a river basin or an administrative unit like a municipality, province, state or nation.

Water footprint within a nation – Is defined as the total freshwater volume consumed or polluted within the territory of the nation.

Water neutral – A process, product, consumer, community or business is water neutral when: (i) its water footprint has been reduced where possible, particularly in places with a high degree of water scarcity or pollution; and

(ii) when the negative environmental, social and economic externalities of the remaining water footprint have been offset (compensated). In some particular cases, when interference with the water cycle can be completely avoided – for example, by full water recycling and zero waste – 'water neutral' means that the water footprint is nullified; in other cases, such as in the case of crop growth, the water footprint cannot be nullified. Therefore 'water neutral' does not necessarily mean that the water footprint is brought down to zero, but that it is reduced as much as possible and that the negative economic, social and environmental externalities of the remaining water footprint are fully compensated.

Water pollution level – Degree of pollution of the run-off flow, measured as the fraction of the waste assimilation capacity of runoff actually consumed. A water pollution level of 100 per cent means the waste assimilation capacity of the run-off flow has been fully consumed.

Water productivity – Product units produced per unit of water consumption or pollution. Water productivity (product units/m^3) is the inverse of the water footprint (m^3/product unit). Blue water productivity refers to the product units obtained per cubic metre of blue water consumed. Green water productivity refers to the product units obtained per cubic metre of green water consumed. Grey water productivity refers to the product units obtained per cubic metre of grey water produced. The term 'water productivity' is a similar term as the terms labour productivity or land productivity, but now production is divided over the water input. When water productivity is measured in monetary output instead of physical output per unit of water, one can speak about 'economic water productivity'.

Water scarcity – See 'blue water scarcity' and 'green water scarcity'.

Water self-sufficiency versus water dependency of a nation - The 'water self-sufficiency' of a nation is defined as the ratio of the internal to the total water footprint of national consumption. It denotes the degree to which the nation supplies the water needed for the production of the domestic demand for goods and services. Self-sufficiency is 100 per cent if all the water needed is available and indeed taken from within the nation's own territory. Water self-sufficiency approaches zero if the demand for goods and services in a nation is largely met with virtual-water imports. Nations with import of virtual water depend, *de facto*, on the water resources available in other parts of the world. The 'virtual-water import dependency' of a nation is defined as the ratio of the external to the total water footprint of national consumption.

Water withdrawal – The volume of freshwater abstraction from surface or groundwater. Part of the freshwater withdrawal will evaporate, another part will return to the catchment where it was withdrawn and yet another part may return to another catchment or the sea.

Index

agricultural policy 110, 111, 113
AQUACROP model 42,132
assimilation capacity 16, 21, 32–34, 36, 66, 76, 78, 86, 91, 148

benchmarking 9, 16, 91–92, 107–108, 157, 160
biodiversity 16, 78–79, 81, 89, 111, 116, 188, 192
blue water
  availability 16, 76, 82–85, 151, 159, 187
  footprint 2–3, 20–21, 23–29, 40–43, 64, 120–121, 141, 145–146, 163–164, 187
  footprint impact index 95, 126–127, 187
  footprint reduction 99–101, 106–109, 159–160
  scarcity 82–86, 94–96, 187
  sustainability of blue water footprint 73–79, 82–96, 153

carbon
  footprint 10–11, 14, 103, 105, 112, 124–125, 127, 155, 160, 166
  neutral 103–104
  offsetting 103, 157, 160
certification 9, 108, 157
comparative advantage 111
consumption pattern 12,99,101, 106, 156
corporate social responsibility 110, 155
cotton 1–2, 22, 47, 106, 123, 155–156, 161
critical load 33–34, 188
crop water requirement 42, 131–133, 135–136, 141, 188
CROPWAT model 42, 44, 131–141
crop yield 41, 44, 131–132, 141, 188

desalination 19, 160
development cooperation 112–113
dilution factor 31, 35, 188

eco-efficient 101
ecological footprint 10, 73, 124–125, 127, 166
effective precipitation 131–133, 188
efficiency of water use
  allocation efficiency 77, 88, 111
  in irrigation 91, 107, 109, 140
  in water-rich areas 74, 102, 159
  productive efficiency 77, 88, 111
effluent
  dilution 66, 147–148
  standards 32, 66
energy
  bioenergy 1, 12, 112, 122, 127, 158–159
  embodied energy 124–125
  energy analysis 10, 124–125
  energy policy 17, 110, 112–113, 157

solar energy 19, 101
water footprint of energy 11, 25, 64
wind energy 101
environmental flow requirements 73, 77–78, 82–86, 121, 151–153, 161, 188
environmental green water requirements 77, 79–81, 121, 188
environmental policy 110–111, 113
EPIC model 42
equitable water use 3, 105, 108–110, 127, 155, 158, 162, 195
evapotranspiration 19–20, 27, 30, 41–44, 79–82, 131–138, 141, 189

foreign policy 110, 112–113

greenhouse gas emissions 14, 38, 124, 127, 166
greenwashing 105
green water
availability 16, 79–81, 189
footprint 2–3, 10, 20, 29–30, 40–43, 64, 127, 141, 146, 163–164, 189
footprint impact index 95, 189
footprint reduction 100–102, 107, 109, 158–160
scarcity 10, 80–81, 95, 158, 190
sustainability of green water footprint 73–82, 89–96
grey water
footprint 2–3, 10–11, 21, 30–41, 46, 64, 66, 120, 142, 146–149, 163–164, 190
footprint impact index 96, 190
footprint reduction 99–101, 106–109, 159
sustainability of grey water footprint 74, 86–87, 90, 96
groundwater
decline 79, 84–85, 108, 121
fossil 26, 85, 120–121, 164
renewable 26, 84, 120–121, 164

hotspot 12, 16, 74, 76, 78–79, 82, 86–90, 92–93, 102–103, 108, 111–112, 149, 157, 190

integrated river basin management 117
integrated water resources management 117
input-output analysis 125
investors 99, 109–110
inter-basin water transfer 29, 54
irrigation
centre-pivot irrigation 134, 138
deficit irrigation 100, 107, 109
drip irrigation 100, 107, 132
furrow irrigation 107
irrigation maps 45
irrigation requirement 131–132, 134, 141, 191
irrigation schedule 42, 44, 109, 133, 135, 138–141
micro-irrigation 134, 138
sprinkler irrigation 100, 107, 132, 134, 138
supplementary irrigation 109
water footprint of irrigation 42–43, 135, 138–141

labelling of products 108, 157, 161
lake level decline 79, 84–85, 108, 121
leaching 37, 41, 45, 100, 109, 142
life cycle assessment 10–11, 70, 94, 96, 125–126, 166

material flow analysis 25
meat 12, 22, 47, 98, 101, 106, 123, 144, 156, 159

offsetting
carbon 103, 157, 160
water 35–36, 103–105, 108, 157–158, 195–196
opportunity cost 10, 30, 77, 88
organic farming 100–101, 109

people planet profit 115
polluter pays principle 87
primary impacts 15–16, 73–74, 76, 78, 88–89, 121, 192
production system 10, 22, 47–49, 65, 92–93, 125, 192
product fraction 50–51, 143–145
product tree 10, 47, 143, 192

rainwater harvesting 26
remote sensing 121
renewable resource 25, 163
return flow 3, 20, 26, 66, 121, 192
risk
business 66, 90, 109–110, 117, 155
financial 110
in supply chain 66, 109, 155
physical 109, 117
regulatory 110, 117
reputational 109–110, 117
run-off 19–21, 34, 36, 77, 82–84, 86–87, 151–152, 159, 187

scarcity rent 77, 88
secondary impacts 15–16, 73–74, 76, 78, 88–89, 121, 192
social licence to operate 117
sugar 123, 135–146, 161
sustainability
assessment 4–5, 12, 15–16, 66, 73–98, 121–122, 126, 129–130, 195
criteria 73, 76–93, 121, 192
geographic 15–16, 73–77, 89, 92–93, 102, 189
of consumption 111, 113, 157
of water footprint 4–5, 15–16, 73–98, 121, 127, 155, 157
reporting 108, 124
sustainable procurement 112
thermal pollution 37, 100, 159

trade
policy 110, 112–113
water saving through 54, 60–61, 111, 113, 157, 189, 191
transparency of products 106–108, 110, 112–113, 156–157, 161
triple bottom line 115
truncation 10–11, 47, 119, 129

uncertainties in water footprint assessment 120, 129
user pays principle 87

value fraction 50–51, 143, 145
virtual water
balance 54, 192
content 46, 167, 193
trade, export/import 24, 46, 54–57, 59–62, 111–113, 123, 193, 196

waste assimilation capacity 16, 21, 32–34, 36, 66, 76, 78, 86, 91, 148
wastewater treatment 25, 31, 35, 37, 90, 101, 106, 108, 148, 158–159

water abstraction 20, 27–28, 34–36, 84, 104, 117, 147–148, 193, 197
water appropriation 2, 19–20, 74, 94–95, 115, 126–127, 164, 193
water consumption 1–3, 10, 21, 64, 66, 162, 193
water dependency
  of a business 162
  of a nation 14, 61, 111–112, 196
water disclosure 107–108, 110
water footprint
  accounting 4–6, 9–14, 19–71, 74, 119–122, 126, 129, 194
  assessment 3–5, 7–9, 115–118, 122, 126, 129–130, 194
  compensation 29, 35–36, 88, 104–105, 108, 157, 195–196
  direct 13–14, 22, 24, 52, 58, 63–65, 106, 156, 188, 191
  end-use 65, 108, 188
  external 14, 23–24, 55–56, 59–63, 111–112, 189
  impact indices 94–97, 126–127, 166, 187, 189, 190
  indirect 11, 13–14, 22, 24, 52, 58, 63–65, 106, 156, 190, 192
  internal 14, 22–24, 55–56, 61–62, 111, 190
  of a business 9, 23, 25, 63–71, 97, 106–109, 119, 123, 155–156, 194
  of a consumer 3–4, 7, 14, 22–23, 25, 46, 52, 75, 97–98, 103, 106–107, 156, 194
  of a process 7–8, 22–41, 48–51, 53, 56, 73, 89–91, 129, 135, 141–142, 145–146
  of a product 2, 7–8, 22–23, 25, 46–53, 57–58, 70, 75, 91–96, 119, 122, 126, 143, 146 161, 195
  of national consumption 7, 9, 14, 22–24, 54–61, 111, 122–123, 191, 195
  of national production 195
  offsetting 35–36, 103–105, 108, 157–158, 195–196
  operational 63–66, 68–69, 100, 106, 111, 156, 162, 191
  organizational 63, 194
  overhead 64–65, 69, 191
  reduction 28, 31, 66, 73, 76, 90, 92–93, 98–113, 116, 121, 129, 155–162, 166
  supply-chain 63–66, 69–70, 106, 156, 162, 192
  sustainability assessment 4–5, 12, 15–16, 66, 73–98, 121–122, 126, 129–130, 195
  within a catchment 7, 62, 76–77
  within a geographically delineated area 7–8, 21–23, 25, 53, 195
  within a nation 7, 9, 14, 22, 24, 54, 56, 59, 191, 195
water neutral 103–105, 195–196
water policy 110–111, 113
water pollution
  diffuse sources 31, 34, 37–38
  point sources 34–36
  water pollution level 78–79, 86–88, 94–96, 102, 148–149, 196
water price 88, 110–113, 158
water pricing protocol 112
water productivity 91, 100–102, 107, 109, 120, 158–160, 189, 196
water quality
  actual concentration 32, 34, 45, 147
  ambient water quality standards 2, 9, 30–40, 45, 66, 73, 77–78,

86–87, 108, 142, 161, 187, 190
effluent standards 32, 66
natural concentration 30, 32–36, 39, 41, 45, 142, 147–148, 191
water recycling 28, 35, 99–101, 104, 106, 108, 156, 159–160, 196
water reuse 26, 28, 35, 66, 91
water saving
in agriculture 100, 107, 109
in industry 108, 156
in households 106, 156
global water saving through trade 60–61, 189
national water saving through trade 60, 111, 113, 157, 191
water scarcity 4, 10, 74–76, 78–88, 91, 94–96, 101–102, 116–117, 127, 155, 160, 187, 190
water self-sufficiency 61, 196
water withdrawal 2–3, 26–27, 51, 54, 65–66, 74, 86, 117, 123, 162, 197